CHALLENGES FOR THE CHEMICAL SCIENCES IN THE 21ST CENTURY

ENERGY AND TRANSPORTATION

ORGANIZING COMMITTEE FOR THE WORKSHOP
ON ENERGY AND TRANSPORTATION

COMMITTEE ON CHALLENGES FOR THE CHEMICAL SCIENCES
IN THE 21ST CENTURY

BOARD ON CHEMICAL SCIENCES AND TECHNOLOGY

DIVISION ON EARTH AND LIFE STUDIES

NATIONAL RESEARCH COUNCIL
OF THE NATIONAL ACADEMIES

THE NATIONAL ACADEMY PRESS
Washington, D.C.
www.nap.edu

THE NATIONAL ACADEMIES PRESS · 500 Fifth Street, N.W. · Washington, DC 20001

NOTICE: The project that is the subject of this report was approved by the Governing Board of the National Research Council, whose members are drawn from the councils of the National Academy of Sciences, the National Academy of Engineering, and the Institute of Medicine. The members of the committee responsible for the report were chosen for their special competences and with regard for appropriate balance.

Support for this study was provided by the National Research Council, the U.S. Department of Energy (DE-AT-01-EE41424, BES DE-FG-02-00ER15040, and DE-AT01-03ER15386), the National Science Foundation (CTS-9908440), the Defense Advanced Research Projects Agency (DOD MDA972-01-M-0001), the U.S. Environmental Protection Agency (R82823301), the American Chemical Society, the American Institute of Chemical Engineers, the Camille and Henry Dreyfus Foundation, Inc. (SG00-093), the National Institute of Standards and Technology (NA1341-01-2-1070 and 43NANB010995), and the National Institutes of Health (NCI-N01-OD-4-2139, and NIGMS-N01-OD-4-2139), and the chemical industry. Any opinions, findings, conclusions, or recommendations expressed in this publication are those of the authors and do not necessarily reflect the views of the organization or agencies that provide support for the project.

International Standard Book Number 0-309-08741-4 (Book)
International Standard Book Number 0-309-52684-1 (PDF)

Additional copies of this report are available from:

The National Academies Press
500 5th Street, N.W.
Lockbox 285
Washington, DC 20055
800-624-6242
202-334-3313 (in the Washington Metropolitan Area)
http://www.nap.edu

Copyright 2003 by the National Academy of Sciences. All rights reserved.

Printed in the United States of America

THE NATIONAL ACADEMIES
Advisers to the Nation on Science, Engineering, and Medicine

The **National Academy of Sciences** is a private, nonprofit, self-perpetuating society of distinguished scholars engaged in scientific and engineering research, dedicated to the furtherance of science and technology and to their use for the general welfare. Upon the authority of the charter granted to it by the Congress in 1863, the Academy has a mandate that requires it to advise the federal government on scientific and technical matters. Dr. Bruce M. Alberts is president of the National Academy of Sciences.

The **National Academy of Engineering** was established in 1964, under the charter of the National Academy of Sciences, as a parallel organization of outstanding engineers. It is autonomous in its administration and in the selection of its members, sharing with the National Academy of Sciences the responsibility for advising the federal government. The National Academy of Engineering also sponsors engineering programs aimed at meeting national needs, encourages education and research, and recognizes the superior achievements of engineers. Dr. Wm. A. Wulf is president of the National Academy of Engineering.

The **Institute of Medicine** was established in 1970 by the National Academy of Sciences to secure the services of eminent members of appropriate professions in the examination of policy matters pertaining to the health of the public. The Institute acts under the responsibility given to the National Academy of Sciences by its congressional charter to be an adviser to the federal government and, upon its own initiative, to identify issues of medical care, research, and education. Dr. Harvey V. Fineberg is president of the Institute of Medicine.

The **National Research Council** was organized by the National Academy of Sciences in 1916 to associate the broad community of science and technology with the Academy's purposes of furthering knowledge and advising the federal government. Functioning in accordance with general policies determined by the Academy, the Council has become the principal operating agency of both the National Academy of Sciences and the National Academy of Engineering in providing services to the government, the public, and the scientific and engineering communities. The Council is administered jointly by both Academies and the Institute of Medicine. Dr. Bruce M. Alberts and Dr. Wm. A. Wulf are chair and vice chair, respectively, of the National Research Council.

www.national-academies.org

ORGANIZING COMMITTEE FOR THE WORKSHOP ON ENERGY AND TRANSPORTATION

ALLEN J. BARD, University of Texas, Co-chair
MICHAEL P. RAMAGE, ExxonMobil, Co-chair
JOSEPH G. GORDON, IBM Almaden Research Center
ARTHUR J. NOZIK, National Renewable Energy Laboratory
RICHARD R. SCHROCK, Massachusetts Institute of Technology
ELLEN B. STECHEL, Ford Motor Company

Staff

JENNIFER J. JACKIW, Program Officer
CHRISTOPHER K. MURPHY, Program Officer
SYBIL A. PAIGE, Administrative Associate
DOUGLAS J. RABER, Senior Scholar
DAVID C. RASMUSSEN, Program Assistant
ERIC L. SHIPP, Postdoctoral Associate
DOROTHY ZOLANDZ, Director

COMMITTEE ON CHALLENGES FOR THE CHEMICAL SCIENCES IN THE 21ST CENTURY

RONALD BRESLOW, Columbia University, Co-chair
MATTHEW V. TIRRELL, University of California at Santa Barbara, Co-chair
MARK A. BARTEAU, University of Delaware
JACQUELINE K. BARTON, California Institute of Technology
CAROLYN R. BERTOZZI, University of California at Berkeley
ROBERT A. BROWN, Massachusetts Institute of Technology
ALICE P. GAST,[1] Stanford University
IGNACIO E. GROSSMANN, Carnegie Mellon University
JAMES M. MEYER,[2] DuPont Co.
ROYCE W. MURRAY, University of North Carolina at Chapel Hill
PAUL J. REIDER, Amgen, Inc.
WILLIAM R. ROUSH, University of Michigan
MICHAEL L. SHULER, Cornell University
JEFFREY J. SIIROLA, Eastman Chemical Company
GEORGE M. WHITESIDES, Harvard University
PETER G. WOLYNES, University of California, San Diego
RICHARD N. ZARE, Stanford University

Staff

JENNIFER J. JACKIW, Program Officer
CHRISTOPHER K. MURPHY, Program Officer
SYBIL A. PAIGE, Administrative Associate
DOUGLAS J. RABER, Senior Scholar
DAVID C. RASMUSSEN, Program Assistant
ERIC L. SHIPP, Postdoctoral Associate
DOROTHY ZOLANDZ, Director

---

[1] Committee member until July 2001; subsequently Board on Chemical Sciences and Technology (BCST) liaison to the committee in her role as BCST co-chair.
[2] Committee member until March 2002, following his retirement from DuPont.

BOARD ON CHEMICAL SCIENCES AND TECHNOLOGY

KENNETH RAYMOND, University of California, Berkeley, Co-chair
ALICE P. GAST, Massachusetts Institute of Technology, Co-chair
ARTHUR I. BIENENSTOCK, Stanford University
A. WELFORD CASTLEMAN, JR., Pennsylvania State University
THOMAS M. CONNELLY, JR., DuPont Company
JOSEPH M. DESIMONE, University of North Carolina, Chapel Hill, and North Carolina State University
CATHERINE FENSELAU, University of Maryland
JON FRANKLIN, University of Maryland
RICHARD M. GROSS, Dow Chemical Company
NANCY B. JACKSON, Sandia National Laboratory
SANGTAE KIM, Eli Lilly and Company
WILLIAM KLEMPERER, Harvard University
THOMAS J. MEYER, Los Alamos National Laboratory
PAUL J. REIDER, Amgen, Inc.
LYNN F. SCHNEEMEYER, Bell Laboratories
JEFFREY J. SIIROLA, Eastman Chemical Company
ARNOLD F. STANCELL, Georgia Institute of Technology
ROBERT M. SUSSMAN, Latham & Watkins
JOHN C. TULLY, Yale University
CHI-HUEY WONG, Scripps Research Institute
STEVEN W. YATES, University of Kentucky

Staff

JENNIFER J. JACKIW, Program Officer
CHRISTOPHER K. MURPHY, Program Officer
SYBIL A. PAIGE, Administrative Associate
DOUGLAS J. RABER, Senior Scholar
DAVID C. RASMUSSEN, Program Assistant
ERIC L. SHIPP, Postdoctoral Associate
DOROTHY ZOLANDZ, Director

Preface

The Workshop on Energy and Transportation took place on January 7-9, 2002, in Washington, DC. This workshop was the second in a series of workshops that comprise the study Challenges in the Chemical Sciences in the 21st Century. The purpose of this study was to carry out a survey of the current status of the chemical sciences, including chemistry and chemical engineering, and its interfaces with other disciplines such as mechanical engineering, physics, materials science, and the biological sciences. The goal of the study was to assess trends across these broad fields and identify key opportunities and challenges.

The Workshop on Energy and Transportation had as its specific focus the contribution that the chemical sciences could make in the development of both emerging and traditional energy sources as well as new and improved transportation. The workshop was attended by approximately 100 individuals from industry, academia, and the federal government with expertise in the chemical sciences. A series of presenters covered issues related to use of the chemical sciences to meet challenges in the areas of energy and transportation. The ideas and challenges identified in the presentations were used as a starting point for breakout sessions, where the participants further developed concepts and needs and identified opportunities. All of this material, from both the presentations and the breakout sessions, were used by the committee as the basis for this report.

The summaries of the presentations in this report contain the opinions expressed by the speakers. Sidebars are included in these presentation summaries to highlight topics of interest related to the presentations. These sidebars were authored by the committee and were based on feedback received by the committee by the workshop participants. The final chapter of this report contains challenges identified by the committee. These challenges are based on the presentations and the feedback from the workshop participants. While the committee relied on this feedback to identify the challenges, sole responsibility for these statements rests with the organizing committee.

>Allen J. Bard
>Michael P. Ramage
>Co-chairs,
>Organizing Committee for the Workshop
>on Energy and Transportation

Acknowledgment of Reviewers

This report has been reviewed in draft form by individuals chosen for their diverse perspectives and technical expertise, in accordance with procedures approved by the National Research Council's (NRC's) Report Review Committee. The purpose of this independent review is to provide candid and critical comments that will assist the institution in making the published report as sound as possible and to ensure that the report meets institutional standards for objectivity, evidence, and responsiveness to the study charge. The review comments and draft manuscript remain confidential to protect the integrity of the deliberative process. We wish to thank the following individuals for their participation in the review of this report:

Rod Borup, Los Alamos National Laboratory
W. Nicholas Delgass, Purdue University
Julie L. d'Itri, U.S. Department of Energy
Theodore H. Geballe, Stanford University
Norman A. Gjostein, Ford Motor Company (retired)
Roy Gordon, Harvard University
David L. Morrison, North Carolina State University
Christine S. Sloane, General Motors Corporation

Although the reviewers listed above have provided many constructive comments and suggestions, they were not asked to endorse the conclusions or recommendations nor did they see the final draft of the report before its release. The review of this report was overseen by Dr. L. Louis Hegedus, Atofina Chemicals,

Inc. Appointed by the National Research Council, he was responsible for making certain that an independent examination of this report was carried out in accordance with institutional procedures and that all review comments were carefully considered. Responsibility for the final content of this report rests entirely with the authoring committee and the institution.

Contents

Executive Summary		1
1	Introduction	11
2	Research Opportunities and Challenges in the Energy Sector	13
3	Fuel Cell Development—Managing the Interfaces	18
4	Interface Challenges and Opportunities in Energy and Transportation	23
5	R&D Challenges in the Chemical Sciences to Enable Widespread Utilization of Renewable Energy	33
6	Nano- and Microscale Approaches to Energy Storage and Corrosion	40
7	Challenges for the Chemical Sciences in the 21st Century	45
8	A Renaissance for Nuclear Power?	49
9	Materials Technologies for Future Vehicles	56
10	Could Carbon Sequestration Solve the Problem of Global Warming?	62
11	The Hydrogen Fuel Infrastructure for Fuel Cell Vehicles	66

12	Opportunities for Catalysis Research in Energy and Transportation	70
13	Role of 21st Century Chemistry in Transportation and Energy	76
14	Future Challenges for the Chemical Sciences in Energy and Transportation	80

Appendixes
	A	Statement of Task	93
	B	Biographies of the Organizing Committee Members	94
	C	Workshop Participants	97
	D	Workshop Agenda	100
	E	Results from Breakout Sessions	104

Executive Summary

Safe, secure, clean, and affordable energy and transportation are essential to the social and economic vitality of the world. As we look to the future—the next 50 years and beyond—there will be many severe challenges for the energy and transportation systems of the world that must be met. The drivers will be created by population growth, economic growth, ever tightening environmental constraints, increasing climate change issues and pressure for limits on carbon dioxide emissions, geopolitical impacts on energy availability and the energy marketplace, a changing energy resource base, and a need for low emissions transportation. Science and technology—specifically chemistry and chemical engineering—will play critical, unique, and exciting roles in enabling the world to meet these challenges.

As the second of six workshops held by the National Research Council for the study Challenges for the Chemical Sciences in the 21st Century, the Workshop on Energy and Transportation sought to identify these key opportunities and challenges for the chemical sciences in the energy and transportation sectors. The workshop featured 12 keynote speakers who addressed the wide spectrum of challenges facing the chemical sciences in energy and transportation. Approximately 100 chemists and chemical engineers from across the industrial, academic, and government research communities attended the workshop (participants are listed in Appendix E). The participants identified key challenges through a series of breakout sessions. The speakers and their presentation titles are listed below.

Patricia A. Baisden　　*Lawrence Livermore National Laboratory*,
　　　　　　　　　　　"A Renaissance for Nuclear Power?

Thomas R. Baker	*Los Alamos National Laboratory,* "Opportunities for Catalysis Research in Energy and Transportation"
Alexis T. Bell	*University of California, Berkeley,* "Research Opportunities and Challenges in the Energy Sector"
Jiri Janata	*Georgia Institute of Technology,* "Role of 21st Century Chemistry in Transportation and Energy"
James R. Katzer	*ExxonMobil,* "Interface Challenges and Opportunities in Energy and Transportation"
Nathan S. Lewis	*California Institute of Technology,* "R&D Challenges in the Chemical Sciences to Enable Widespread Utilization of Renewable Energy
Ralph P. Overend	*National Renewable Energy Laboratory,* "Challenges for the Chemical Sciences in the 21st Century"
Stephen W. Pacala	*Princeton University,* "Could Carbon Sequestration Solve the Problem of Global Warming?"
Venki Raman	*Air Products and Chemicals,* "The Hydrogen Fuel Infrastructure for Fuel Cell Vehicles"
Kathleen C. Taylor and Anil Sachdev	*General Motors Corporation,* "Materials Technologies for Future Vehicles"
John R. Wallace	*Ford Motor Company,* "Fuel Cell Development—Managing the Interfaces"
Henry S. White	*University of Utah,* "Nano- and Microscale Approaches to Energy Storage and Corrosion"

Summaries of these presentations are presented in the main body of the report. A full agenda for the workshop may be found in Appendix D. The organizing committee integrated the input from the presentations and breakout sessions to develop this report.

In order to define the energy and transportation challenges and opportunities for the chemical sciences in the 21st century, the future needs can be divided into two time frames—midterm (through 2025) and long term (2050 and beyond).[1] While these scenarios can be debated, the drives they create in the chemical sciences are not greatly affected by the severity of the scenarios. They do point to a

---

[1]These future needs were identified by the committee based on the Workshop presentations. For each need, the presentation from which it was identified is given in parentheses.

need to enhance the energy efficiency of fossil fuels in production and utilization, to develop a diverse set of new and carbon-neutral energy sources for the future, and to maintain a robust basic research program in the chemical sciences so that the technical breakthroughs will happen to enable this future. In the midterm:

- World energy demand will increase approximately 50 percent above 2002 levels. (Alexis Bell)
- Fossil fuels will remain abundant and available as well as continue to provide most of the world's energy. (Nathan Lewis)
- There will be a drive toward fuels with higher hydrogen/carbon ratio, but balanced against the need to utilize the extensive low hydrogen/carbon coal resource base in the United States. (Venki Raman, Nathan Lewis)
- Tighter environmental constraints will be imposed. (Nathan Lewis)
- Government-mandated carbon dioxide limits will be initiated, leading to a need for carbon dioxide sequestration technology and/or the introduction of large amounts of carbon-neutral energy. (Stephen Pacala)
- A real but limited role will be found for wind and hydro energy sources. (Nathan Lewis)
- Nuclear, solar, and biomass energy will play a growing role in the nation's energy mix. (Patricia Baisden, Jiri Janata)
- Cost-effective hydrogen fuel cell technology for transportation and power will be developed. (John Wallace, James Katzer)
- A significant penetration of vehicles with new high-efficient clear power sources will be seen in the transportation market. (John Wallace, James Katzer)
- Most hydrogen will be produced from fossil fuels.[2]

In the long-term:

- World energy demand will rise to approximately two and a half times the present energy usage. (Nathan Lewis)
- Fossil fuels like coal and natural gas will remain abundant and available, but a serious limitation on their use will arise because of worldwide constraints on carbon dioxide emissions. (Nathan Lewis, Alexis Bell)
- There will be a need for significant carbon-neutral energy. (Most of the presenters)
- Fully developed carbon dioxide sequestration technology will be one of the important approaches to solving the energy problem. (Stephen Pacala)
- Coal and nuclear energy will continue to play a significant role in meeting world power demands. (Nathan Lewis, Alexis Bell)

---

[2]Venki Raman, in his presentation to the Energy & Transportation Workshop, noted that presently eighty percent of the hydrogen produced is made from natural gas steam methane generation.

- Renewable energy (wind, biomass, geothermal, photovoltaics, and direct photon conversion—for example, solar photovoltaic water splitting) will play an increasingly important role. (Nathan Lewis, Ralph Overend)
- Most of world's vehicles will run on hydrogen from a carbon-free source or other fuels that are carbon-neutral. (John Wallace, James Katzer, Venki Raman)
- New cost-effective solar technology will be widely available. (Nathan Lewis, Ralph Overend)
- Hydrogen and distributed electricity will be produced by solar energy, either through photovoltaic electrolysis or by direct solar photoelectrolysis. (Nathan Lewis, Ralph Overend)

Many of the issues discussed in this report, from increased energy efficiency from fossil fuels, to reduction of pollution, to sequestration of carbon dioxide, to development of new materials for vehicle fabrication, to new low-cost renewable energy technologies, if not wholly chemical in nature, contain significant chemical science content. As chemical scientists seek to address these issues, the crosscutting nature of many of these challenges should be recognized. Many of the challenges in energy and transportation will be met with technologies that have broad applications in a number of different fields. By working with scientists and engineers in other disciplines, such as materials scientists, bioscientists, geologists, electrical engineers, information scientists, mechanical engineers, and others, a multidimensional approach to these challenges will be accomplished—and the likelihood for comprehensive new solutions will increase significantly.

The path will not be straightforward, however, particularly in the United States. Interest in and appreciation of the importance of science and technology are decreasing. Fewer U.S. students are entering technical careers. Energy research is decreasing significantly in both the private and public sectors. While this workshop and report do not address these issues, they must be resolved or the United States will be in jeopardy of not being able to meet its future energy and transportation requirements.

The following challenges were identified resulting from the presentations and discussions at the Workshop. Although these challenges were identified as a result of the Workshop, final responsibility for these statements rests with the organizing committee.

ENERGY

Fossil fuels will remain an abundant and affordable energy resource well into the 21st century. Since potential limitations on carbon dioxide emissions may restrict their utilization in the long term, it is imperative that chemical sciences research and engineering focus on making significant

increases in the energy efficiency and chemical specificity of fossil fuel utilization.[3]

Professor Bell identified new multifunctional highly selective catalysts and membranes and corresponding process technologies as the key research areas where opportunities will exist for major steps forward. These new catalysts and materials will allow much greater process efficiency (reduced carbon dioxide) through operations at lower temperatures and pressures and also by combining multiple process functions (i.e., shape selectivity and oxidation) in a single catalyst particle, thus reducing the number of process units in a plant.

These new materials and processes will increase the efficiency and environmental cleanliness of hydrocarbon production and refining and also enable refineries to produce chemically designed fuels for future vehicle power trains. These chemically designed fuels will play a key role in new power trains. These engines will require fuels that can optimize the efficiency of the entire power cycle while at the same time produce essentially no harmful exhaust. The best way to accomplish this is by designing the engine and fuel interactively, and this will lead to more chemical specificity requirements on the fuel.

Natural gas has tremendous potential for meeting the energy needs of the future because it has a high hydrogen/carbon ratio and can be converted to H_2 and environmentally clean liquid fuels.[4]

Current technology for converting natural gas to liquid fuels is by Fischer-Tropsch Technology, which converts methane to syngas (CO and H_2) and the syngas to liquids (the Fischer-Tropsch step). While there have been major advances in the technology in the past decade, it is much less energy efficient than today's refining processes. New catalysts, membranes, and processes are needed which will convert methane directly to H_2 and liquid fuels without going through syngas. This would tremendously increase the energy efficiency of methane conversion. Liquid products from these processes are chemically pure, containing no heteroatoms (i.e., sulfur, nitrogen, metals)

Management of atmospheric carbon dioxide levels will require sequestration of carbon dioxide. Research and development into methods to cost effectively capture and geologically sequester carbon dioxide is required in the next 10 to 20 years.[5]

---

[3]Alexis T. Bell, University of California, Berkeley, presentation at the Workshop on Energy and Transportation.

[4]Alexis T. Bell, University of California, Berkeley, Nathan Lewis, California Institute of Technology, presentations at the Workshop on Energy and Transportation.

[5]Stephen W. Pacala, Princeton University, presentation at the Workshop on Energy and Transportation.

As noted in Professor Pacala's presentation, effective management of the increasing anthropogenic output of carbon dioxide into the atmosphere will be a significant challenge for the chemical sciences and engineering over the next century. Development of sequestration technology to address this issue will require a thorough understanding of carbon dioxide chemistry and geochemistry along with an elaboration of the mechanisms involved in carbon dioxide absorption, adsorption, and gas separation. Also, effective sequestration will require new engineering knowledge to capture and transport the carbon dioxide to the sequestration site—most likely a geological reservoir. A more thorough understanding of the geochemical, geological, and geophysical nature of the sequestration site will be required to ensure that carbon dioxide does not escape over centuries of storage.

Biomass has the potential to provide appreciable levels of fuels and electric power, but an exceptionally large increase in field efficiency[6] is needed to realize the huge potential of energy from biomass.[7]

Biologically based strategies for providing renewable energy can be grouped into two major categories: (1) those that use features of biological systems to convert sunlight into useful forms (e.g., power, fuels) but do not involve whole living plants, and (2) those involving growth of plants and processing of plant components into fuels and/or power. Both are very important. Long-term improvements can be expected in the development of both biomass resources and the conversion technologies required to produce electric power, fuels, chemicals, materials, and other bio-based products. As molecular genetics matures over the next several decades, for example, its application to biomass energy resources can be expected to significantly improve the economics of all forms of bio-energy. Improvements in economics, in turn, will likely lead to increased efforts to develop new technologies for the integrated production of ethanol, electricity, and chemical products from specialized biomass resources. Near-term markets exist for corn-ethanol and the co-firing of coal-fired power plants.

By the middle of the 21st century, global energy consumption will more than double from the present rate. To meet this demand under potential worldwide limits on carbon dioxide emissions, cost-effective solar energy must be developed.[8]

---

[6]In agriculture, field efficiency is the ratio of effective field capacity and theoretical field capacity.

[7]Alexis T. Bell, University of California, Berkeley, Nathan Lewis, California Institute of Technology, Ralph P. Overend, National Renewable Energy Laboratory, presentations at the Workshop on Energy and Transportation.

[8]Alexis T. Bell, University of California, Berkeley, Nathan Lewis, California Institute of Technology, Ralph P. Overend, National Renewable Energy Laboratory, presentations at the Workshop on Energy and Transportation.

EXECUTIVE SUMMARY

At present consumption levels, the supply of carbon-based fuels will be sufficient to meet our energy needs for well over a century. However, as noted in both Professor Bell's and Professor Lewis' presentations, the anticipated growth in energy demand over the next century, combined with climate change concerns, will drive the increased use of alternative sources of carbon-neutral energy. While a number of potential sources of renewable energy show promise for meeting part of this increased demand, including wind, biomass, geothermal, and expanded use of hydroelectric sources, cost-effective solar power will likely be required to meet the largest portion of this demand. However, in order for use of solar power to increase substantially over the 21st century, new discoveries in photovoltaic and photochemical energy technologies must be made to reduce costs, increase conversion efficiency, and extend operating life. Advanced materials such as organic semiconductors and semiconducting polymers are needed to reduce energy costs from photovoltaics and make them competitive for electric power and hydrogen generation. Current silicon-based photovoltaics are highly efficient, but also very expensive. New technologies are needed to bring costs down. New photovoltaic materials and structures with very low cost-to-efficiency ratios are needed to produce a step change in the use of photovoltaic technology. For example, the use of grain boundary passivation with polycrystalline semiconductor materials might lead to the replacement of expensive single crystal-based technologies. The development of new, inexpensive, and durable materials for photoelectrochemical systems for direct production of hydrogen and electricity generation will be one of the main factors that will enable broad application of solar power to meet future energy needs.

Widespread use of new, renewable, and carbon-neutral energy sources will require major breakthroughs in energy storage technologies.[9]

Development of these technologies is dependent, in part, on breakthroughs in the design of energy storage systems due to the intermittent nature of many forms of renewables. Batteries, whose basic design has remained relatively unchanged for over a century, need to be fundamentally reexamined, as they will play an important role in meeting future energy needs. For example, advances in nanotechnology and its use in three-dimensional electrochemical cells offer the possibility of increased energy density compared to conventional batteries, but these advances are still in the early stages of development. In addition, fundamental research breakthroughs are needed on thin-film electrolytes in order to develop high-power-density batteries and fuel cells.[10]

---

[9]Nathan Lewis, California Institute of Technology, Henry S. White, University of Utah, presentations to the Workshop on Energy and Transportation.

[10]At present fuel cell systems are being piloted for distributed generation backup power. This may provide another source of energy storage.

For full public acceptance of nuclear power, issues such as waste disposal, reactor safety, economics, and nonproliferation must be addressed.[11]

Future energy consumption trends indicate the need for additional sources of carbon-neutral energy. No one source of power will be sufficient to meet all of the projected increase in future power needs. Dr. Baisden in her presentation noted that nuclear power offers a plentiful supply of energy that is free from local emissions and produces no carbon-based greenhouse gases. However, nuclear power is unique in that political considerations are as important as technical challenges. One of the main technical challenges is waste management and disposal. Significant amounts of uranium can be reprocessed and reused in reactors, but this technology comes with significant concerns about nuclear proliferation and safety. Particularly in light of recent terrorist actions in the United States, the development of safe nuclear waste forms that not only will survive long-term repository storage but also allow secure transit to a repository remains an important priority.

Another significant issue facing the U.S. is the growing shortage of nuclear technical expertise. This threatens the management of the United States' currently installed nuclear capacity and certainly the development of the science and engineering needed to expand nuclear energy use in the future. The training situation is dire in nuclear chemistry, radiochemistry, and nuclear engineering. To address this shortage reinvestment in the education system will be required.

TRANSPORTATION

Vehicle mass reduction, changes in basic vehicle architecture, and improvements in power trains are key to improved vehicle efficiency. The development and use of new materials are crucial to improved fuel efficiency.[12]

Dr. Sachdev noted in his presentation that reductions in the body mass of passenger vehicles will depend to a great extent on the successful integration of new lightweight materials. The dual needs in these applications—for materials that are both lightweight and strong—continue to present challenges and opportunities in the chemical sciences.

The development of new polymers and nanocomposite materials will play an increasing role in vehicle mass reduction. The combination of high strength and light weight makes them ideal for many of these applications. Along with new

---

[11]Patricia A. Baisden, Lawrence Livermore National Laboratory, Jiri Janata, Georgia Institute of Technology, presentations to the Workshop on Energy and Transportation.

[12]James R. Katzer, ExxonMobil, Kathleen C. Taylor and Anil Sachdev, General Motors Corporation, presentations to the Workshop on Energy and Transportation.

materials, manufacturing and recycling processes will have to be developed that are both cost-effective and environmentally effective.

As with the development of new catalysts, effective new materials benefit from a thorough understanding of structure/property relationships. This involves multiscale modeling and experimental efforts in surface science, including morphology. Enabling the use of new materials will also require extensive development of new nano- and microfabrication techniques, including biodirected or self-assembly syntheses.

Cost remains one of the main factors that determine both the need and the acceptance of new materials for applications in energy and transportation. In addition, passenger safety, which may be affected by the development of more lightweight vehicles, must also be taken into consideration. The imperative of low-cost, high-performance materials in the automotive industry will be driven by future environmental and corporate average fuel economy (CAFE) standards.

Reduced material cost is key to widespread use of the proton exchange membrane (PEM) fuel cell.[13]

As with other materials challenges, selective and energy-efficient separations are a highly desirable characteristic in many areas of energy and transportation research and engineering. Development of low-temperature, corrosion-resistant, thin membranes will further PEM development. However, development of new catalytic materials to replace the very expensive platinum in today's design is the most critical need.[14] Low-cost materials in fuel cells will be one of the key deciding factors in whether the United States readily transitions to a hydrogen economy.

The lack of hydrogen generation, transportation, and storage infrastructure presents one of the main challenges to introducing hydrogen into the mass market as a transportation fuel and energy carrier.[15]

Effective hydrogen management and creation of the needed infrastructure will both be key to widespread adoption of hydrogen fuel cells to meet the nation's energy needs for transportation and power. The challenges are great. New-generation technology is needed in the short- to midterm for hydro-carbon based local refueling sites. In the long term, science and technology will have to be developed to generate hydrogen from carbon-free sources such as water, or at a minimum from carbon neutral sources. Whether this new energy source is based

---

[13]John R. Wallace, Ford Motor Company, presentation to the Workshop on Energy and Transportation.

[14]A complementary goal to replacing expensive Pt in today's design is to develop catalysts with reduced Pt loading.

[15]Venki Raman, Air Products and Chemicals, presentation to the Workshop on Energy and Transportation.

on nuclear, solar, or something that remains undiscovered, it will be one of the largest technical challenges the chemical sciences has ever undertaken.

Another significant challenge to effective hydrogen management is the development of efficient hydrogen storage, both onboard the vehicle and at a hydrogen generation facility. As with many other challenges, effective hydrogen storage is a crosscutting one that will require breakthroughs in a number of research areas. Progress is being made with metal hydrides and carbon nanotubes—but a commercial solution is a long way off. New materials will be key.

These technical challenges regarding hydrogen presently hinder widespread commercial use of hydrogen fuel cell technology for transportation and power. Until these challenges are met, it is unlikely that fuel-cell-powered vehicles will ever make up a significant portion of the passenger vehicle market.

CROSSCUTTING

Development of new, less expensive, more selective chemical catalysts is essential to achieving many challenges in both energy and transportation.

Catalysts are expected to play a role in virtually every challenge where chemical transformations are a key component, as evidenced by the fact that virtually every presenter at the workshop mentioned catalysis. The development of new catalysts to solve challenges in energy and transportation will require the ability to design catalysts for specific needs. Utilization of new materials, nanotechnology, new analytical tools, and advanced understanding of structure/property relationships will create major catalytic advances. One of the major areas where these advances are needed is in controlling nitrogen oxide emissions from lean-burn engines and nitrogen oxide from coal power plants. Others are increased energy efficiency of fossil fuel processes, delivery of chemically designed fuels to new vehicle power systems, and direct conversion of natural gas into liquid fuels and hydrogen. Another is the discovery of less expensive catalysts for the electroreduction of oxygen and the oxidation of fuels that can play an important role in fuel cells. Catalysts for promoting oxygen and hydrogen evolution from water are also important in the design of photoelectrochemical systems.

Because this report is based only on a two-day workshop, details of chemical science research and engineering programs need to be further developed for each finding. The workshop's organizing committee suggests that the National Research Council pursue development of these detailed programs because of the importance of energy and transportation to the nation.

1

Introduction

The U.S. economy and our quality of life are heavily dependent on economical, plentiful, and reliable supplies of energy. The production of enhanced energy-efficient transportation, improvements in existing sources of energy, and the development of new sources of energy were key challenges to both chemistry and chemical engineering for much of the 20th century. Future challenges to energy and transportation will come from growth in world energy demand, population growth, renewable energy, availability of hydrogen as an energy resource, environmental constraints such as climate change, and light-duty vehicle power train advancements. The major issue will always be to meet these challenges with cost-effective technologies.

A world population of 10 billion to 11 billion is projected by 2050, combined with an annual increase in the average world gross domestic product per person of 1.6 percent (historical average). Counter-balanced against these trends is a decrease of percent annually in the energy consumption per unit of GDP because of expected increases in the efficiency of energy utilization. The expected annual global energy consumption in 2050 will be more than twice current levels.

Fossil fuels will continue to be economical and widely available through at least the first half of the 21st century because they are plentiful, readily available, and fundamentally cheap. Currently, 75 million barrels of oil are consumed every day. By 2020 it is anticipated that 112 million to 114 million barrels will be consumed daily. The availability of oil will not be a problem outside political issues that may appear. In fact, in the past 20 years the proven reserves of crude oil have increased by over 50 percent from a little over 600 billion barrels to about 1 trillion barrels.

Renewable energy technologies are expected to grow significantly. However, over the next 20 years, even with this rapid growth, renewable wind and solar energy will contribute only about 1 million barrels of oil equivalent per day energy out of 300 million. Beyond 2020, growth of renewable energy technologies may be very rapid, depending on advances in science and technology.

To meet future energy demands, research on technologies that will help meet future demands on energy and transportation must be pursued today. Many of these technological developments will depend on advances made in the chemical sciences, from the development of more efficient catalysts, to improvements in separation technologies, to the development of new materials for photovoltaic cells.

The following sections summarize presentations given at the workshop on Challenges for the Chemical Sciences in the 21st Century: Energy and Transportation.

2

Research Opportunities and Challenges in the Energy Sector

Alexis T. Bell,
University of California, Berkeley

Energy usage in the United States presently is largely from fossil fuels—petroleum, natural gas, and coal. Petroleum is used to produce transportation and heating fuels and to a smaller degree it is a source of chemicals and lubricants. Natural gas is used to produce electricity along with domestic and industrial heating. Coal is largely used to produce electricity.

In the 21st century this pattern of energy usage probably will not change significantly. The main reason is that carbon-based fuels will remain plentiful and low in cost. Examination of the reserve supply of fossil fuels indicates a substantial amount of petroleum still in the ground and accessible. At present consumption rates, and analyzing only proven reserves for fossil fuels, it is anticipated that oil reserves will last for at least another 40 years, supplies of gas another 70 years, and coal supplies 200 years. If likely reserves that have not yet been discovered are included, fossil fuels will be plentiful for decades to come.

In light of the supply predictions, what are the drivers for changing the nation's primary fuel sources? The first is a desire to reduce the nation's dependence on imported petroleum. The second driver is the need for clean-burning fuels, including gasoline containing less sulfur and diesel fuels that produce less soot, which would reduce the impact of vehicle emissions and other combustion sources on human health. A third and very significant motivator is the increasing concern about man-made carbon dioxide emissions being released into the atmosphere.

The extent to which man-made carbon dioxide emissions contribute to global warming is still an issue of considerable debate. Analysis indicates that while the contribution to the total amount of this gas in the atmosphere from anthropogenic sources amounts to 3 to 4 percent, the problem is that carbon dioxide remains in the atmosphere for a very long time and thus accumulates. If currently projected rates of fuel consumption continue, a doubling of carbon dioxide in the atmosphere is anticipated by the end of this century. Even though the question of whether man-made carbon dioxide emissions are responsible for global warming is still unresolved, there is a growing global consensus that these man-made carbon dioxide emissions must be reduced.

Options for reducing man-made carbon dioxide emissions include improvements in the efficient use of carbon-based fuels. This goal could be achieved by moving to lighter-weight vehicles that consume less fuel and also by switching from gasoline to diesel fuel. With a threefold increase in efficiency, it is estimated that a two and one-half to threefold decrease in the amount of carbon emitted per mile driven would result.

Another means to reduce man-made carbon dioxide emissions is sequestration in the land or the ocean. When carbon dioxide is produced locally it may be possible to efficiently separate it from other gases, concentrate it, and dispose of it. A number of complex scenarios may be envisioned to accomplish this disposal, from pumping it into the ocean, to displacing methane in coal mines, to storage in depleted hydrocarbon reservoirs.

FUELS WITH HIGHER HYDROGEN CONTENT

One of the most promising means to mitigate carbon dioxide emissions is to either reduce the use of carbon-based fuels and/or increase the hydrogen content of the fuels that are used. Looked at historically, it is clear the trend in fuel use over the past 150 years has been toward fuels with progressively higher hydrogen-to-carbon content, starting with wood, moving to coal and petroleum, then to methane. If this trend continues, we will eventually move toward a nonfossil-based hydrogen economy.

Over the past decade, natural gas usage increased by 20 percent, petroleum utilization increased by 12 percent, and coal use decreased by 6 percent. This trend indicates that for the near term at least there is already a noticeable increase in the use of more hydrogen-rich fuels.

Natural Gas

Natural gas has a high hydrogen-to-carbon content and is plentiful worldwide. It can be brought to market in four different ways—pipelines, liquefied natural gas, conversion to electricity, or conversion to liquid products that can be pipelined. It is also possible to envision in the near future a natural gas refinery for which natural gas will serve as the principal feedstock. This type of facility would provide an effective means to use natural gas to produce electricity and liquid fuels while also allowing for the recovery of carbon dioxide on site that then could be sequestered.

One of the ways in which natural gas could be converted to liquid products is by Fischer-Tropsch synthesis. In this process, methane is reformed with steam and oxygen to produce a synthesis gas that is a mixture of carbon monoxide and hydrogen. The synthesis gas is then reacted over a catalyst to produce a variety of fuels. However, recently the most emphasis has been on the production of high-cetane, sulfur-free diesel fuel. Fischer-Tropsch fuels can be produced at the equivalent of $14 to $20 a barrel of oil, and plants with capacities of 10 to 100,000 barrels a day have either been built or designed.[1]

Another liquid fuel that can be produced in a natural gas refinery is methanol. Methanol is itself a high-octane fuel used in racecars, and it is sulfur free. It can be reformed readily to hydrogen and carbon dioxide, and the hydrogen can be used for fuel cells. Direct methanol fuel cells convert methanol into protons, free electrons, and carbon dioxide, thus providing a safe and simple-to-use energy source.[2] The byproducts of this type of fuel cell—steam and carbon dioxide—are produced in such small amounts that these fuel cells are particularly environmentally friendly. However, methanol, while readily degradable, presents problems. It has half the energy content of gasoline on a per-volume basis and has a much higher acute toxicity than gasoline. To use methanol as a common fuel would require modifications in the storage and delivery infrastructure.

An alternative to methanol that possibly may be more attractive is dimethyl ether, which is produced by coupling two methanol molecules. Dimethyl ether provides a good substitute for liquefied petroleum gas. It burns very cleanly, it is a high-cetane sulfur-free smokeless diesel fuel, it can be easily reformulated to hydrogen and carbon dioxide, and it can readily be converted to gasoline using zeolites. Balanced against these benefits is the fact that dimethyl ether is very volatile and, similar to methanol, would require a new storage and delivery infrastructure.

Biomass presents another option for mitigation of carbon dioxide in the atmosphere. The basic concept is to grow plants to provide cellulose that can be converted to ethanol with known technology and then concentrate the ethanol produced up to 100 percent and use it as a fuel. It is estimated that use of biomass could be a carbon dioxide-neutral technology, meaning that as much carbon dioxide is consumed as is returned to the atmosphere. It has been estimated that there is sufficient land mass available in the United States to replace the 1.3×10^{11} gallons per year of gasoline used in 2000[3] without substantially impacting land other-

---

[1] G.N. Choi, S.J. Kramer, S.S. Tram, J.M. Fox III. July 9-11, 1996. "Economics of a Natural Gas Based Fischer-Tropsch Plant." First Joint Power and Fuel Systems Contractors' Conference. Pittsburgh, PA.

[2] One issue to consider with direct methanol fuel cells is their relative efficiency compared to indirect methanol fuel cells. Direct methanol fuel cell anode catalysis is currently poor and requires a high overpotential. When comparing an indirect methanol fuel cell (reforming of methanol to hydrogen and carbon dioxide, feeding the hydrogen to a fuel cell) and a direct methanol fuel cell, the efficiency on an indirect methanol fuel cell system is much higher, thus the carbon dioxide emissions are lower.

[3] "Worried Drivers," ExxonMobil Corporation, Houston, TX, 2001.

wise used for agriculture. As with other fuels, ethanol would require change in the fuel infrastructure.

Hydrogen

Hydrogen in many respects may be the ultimate fuel. It is completely carbon free and sulfur free. It can be used for spark injection engines and can be combined with natural gas and burned in such engines as well. It is an ideal fuel for fuel cells, although problems with storage and distribution would require an entirely new infrastructure.

There are a number of significant issues regarding hydrogen production and bringing it to market. Current technology requires a high water-to-methane ratio to avoid carbon deposition on the catalyst used for the reforming of methane. Electrolysis of water is an alternative, but if carbon-based fuels are used for electricity generation, double the amount of carbon dioxide is produced per mile when compared to gasoline. Photovoltaic generation of hydrogen is attractive, but the silicon technology required for making energy through electrolyzed water is at present prohibitively expensive.

The use of fuel cells running on hydrogen produced by the reforming of a liquid fuel has been much discussed as a fuel-efficient means of powering automobiles. While this technology has been demonstrated, approximately 100 to 150 grams of noble metal—principally platinum—would be required per car using present technology. For reference, this is two orders of magnitude more precious metal per car than is required for the catalytic converter. Since at present roughly one half to one third of the world's supply of precious metals is being put into automobiles for use in automotive conversion technologies, it is hard to envision fuel cells as a widespread alternative for automobiles unless the requirements for precious metals can be reduced by one to two orders of magnitude, or other catalytic approaches are developed that do not require the use of precious metals.

An additional challenge to the use of fuel cells for automobiles is response time. Currently, fuel cells have a response time of 15 seconds from 10 percent power to 90 percent. In order to be viable, this response time must drop to 1 second. Because they require liquid water to operate, a further challenge is to operate fuel cells in subfreezing temperatures. In addition, the current cost per kilowatt-hour for fuel cells must be reduced from $300 down to $45.

OPPORTUNITIES FOR THE CHEMICAL SCIENCES

To meet the challenges discussed above, research opportunities for the chemical sciences may be envisioned in four areas. First, in carbon dioxide sequestration, an understanding is required regarding how molecular carbon dioxide interacts with various species present in coal mines or other geological formations where carbon dioxide might be stored. It is particularly important to understand carbon dioxide flow in permeable underground formations, as well as the geology and geochemistry of these formations, to ensure that large amounts of carbon dioxide do not escape back into the atmosphere.

In order for methane to be used for energy, a number of developments must take place. New catalysts for the conversion of methane to oxygenated products are needed, particularly for transformation of methane to formaldehyde or methanol without first converting methane to syngas. Highly effective catalysts are also required to lower the reaction temperature and increase overall efficiency.

Photovoltaics also require significant research activity in the chemical sciences. Low-cost methods are required for producing solar-grade silicon for photovoltaic cells. Better solar cell materials are needed than the presently utilized amorphous silicon. These materials must be more efficient without the use of heavy metals such as cadmium, tellurium, indium, and lead, which present significant environmental issues. An understanding of the degradation process of photovoltaic cells is needed, as is an answer to why these materials lose their effectiveness after prolonged exposure to the sun. Finally, there is a need to develop catalysts for the efficient photochemical conversion of water.

A fourth area for research opportunities is in the development of fuel cells. There is a need to develop electrode materials for methanol-based fuel cells. This would allow for the use of liquid fuel directly without a reformer. Less expensive alternatives to Nafion are required for fuel cell membranes. Rapid-response onboard reformers are also needed for potential use in conjunction with the fuel cell to convert liquid fuel to hydrogen.

Requirements to Achieve these Goals

In order for the chemical sciences to take advantage of the research opportunities outlined above and to achieve the goals set, two major requirements must be met. First, public policies must be set that signal the need for development and deployment of new technologies in the energy sector that meet four criteria:

1. Technologies that are carbon efficient;
2. Technologies that enable carbon dioxide sequestration;
3. Technologies that enable the use of natural gas to produce liquid fuels; and
4. In the longer term, technologies that enable the use of biomass, solar energy, and renewable sources of energy in general.

A second major requirement is a major commitment toward federal support for energy-related research. In the 13-year period from 1985 to 1998, real dollar U.S. investment in energy research dropped 36 percent. Worldwide there was a 33 percent drop. This commitment should not be a short-term one lasting 1 to 2 years but rather a sustained investment over several decades.

3

Fuel Cell Development—Managing the Interfaces

John R. Wallace,
Ford Motor Company

INTRODUCTION

The development of a commercially viable fuel cell system for transportation has been one of the largest problems that the automobile industry has faced in the past 50 years. Its solution will require the coordination and interaction of many different disciplines. Although the fuel cell stack is fundamentally an electrochemical device, the surrounding air and water management systems interact significantly with the stack design. Fluid dynamics, mechanical engineering, and control systems skills are required to develop a properly optimized system. The interaction of the complete vehicle with the fueling infrastructure adds further to the overall complexity of this emerging technology. Any organization or set of organizations with a goal of successfully commercializing fuel cells in transportation must reflect these interactions and manage these complex interfaces.

Sir William Robert Grove invented the fuel cell or "gas battery" in the 1840s, but the discovery of the "fuel cell effect" by Christian Friedrich Schoenbein dates back to 1838. The first practical fuel cells were not built until the Gemini and Apollo space programs in the 1960s and are still used in space today. The difference between building a successful fuel cell and a commercially successful fuel cell, however, is the same difference between putting a man on the moon and putting 10,000 men on the moon every day at an affordable price. Despite all of the challenges associated with fuel cell technology, vehicles powered by fuel cells promise zero tailpipe emissions with improved fuel efficiency and fuel flexibility. Alternative fuel sources in the transportation technology portfolio are preferred, so that no matter what changes occur in the fuel industry, people and goods can move around. As it currently stands, an enormous industrial pyramid rests on a single transportation fuel.

FUEL CELLS

For vehicles, proton exchange membrane fuel cells are probably the most practical design (Figure 3.1). The most important part of a fuel cell is the membrane, which must be an ion conductor, an electronic insulator, an impermeable gas barrier and also possess good mechanical strength. However, the key issues in making a practical fuel cell are nonelectrochemical. These include the acts of delivering the gases to the fuel cell membrane, removing the water, removing the heat from around the system, and controlling humidity and pressurization of gases. There are still many challenges for electrochemists, chemists, and chemical engineers. For example, a membrane that is more tolerant of environmental conditions for gases of varying pressures will allow for the elimination of various system components, which can be very expensive due to their use of stainless steel. The technical challenge is in fabricating a membrane to be thin enough so that the hydrogen side of the gas supply does not need to be humidified. However, as membranes get thinner, reliability over long periods of time becomes an issue due to faradaic losses. If the membrane is too thick, additional components must be added to humidify the hydrogen.

In a vehicle fuel cell stack, which has over 400 cells in series, the situation is even more complicated. Well over 90 percent of fuel cell industry funds are not spent on the membrane but on moving these gases in and out of the fuel cell stack,

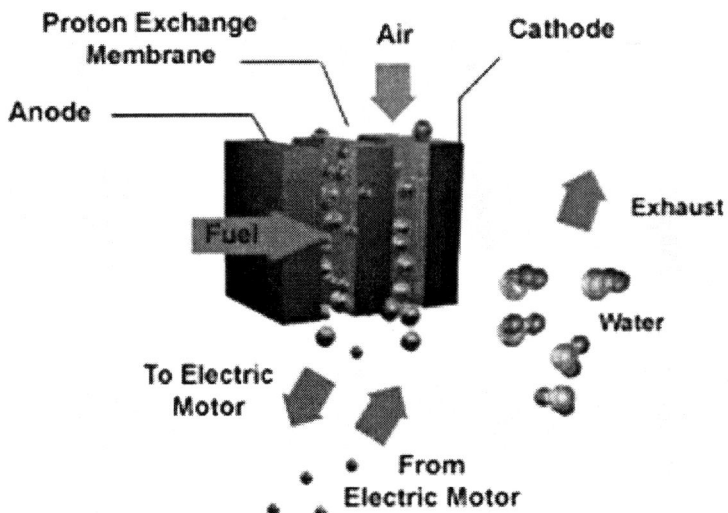

FIGURE 3.1 Configuration of a typical proton exchange membrane (PEM) fuel cell. Source: National Fuel Cell Research Center.

managing the system, and creating the environment where the membrane can do its job. Fuel cell research, however, is mainly performed in a lab where gases are supplied at exactly the right humidity, pressures, and so on. The actual commercial problem, development of a fuel-cell-powered vehicle that has a life of 15 years and 150,000 miles under terrible external environmental conditions, has not been approached. Fuel cell technology may be developed to permit its use for power generation in the near future, but the cost is prohibitive for vehicle transportation.

Tolerances are also not well understood. A fuel cell stack with over 400 cells operating in this environment contains sealant, which is literally miles long. Seals will start to fail after the fuel cell is bumped and jostled on the highway and while temperature shifts between hot and cold, and the cell is turned off and on. With zero tolerance for safety failures, hydrogen leaks cannot occur with these vehicles. Additionally, every cell has to be identical or the system cannot be managed. Unfortunately, that kind of tolerance control is not yet available.

An ideal fuel cell system will have minimal components outside of the stack and will operate using ambient, unhumidified hydrogen. Although fuel cells are very efficient, they do not release much heat through the exhaust. Even though they generate less heat than an internal combustion engine, the system requires the addition of cooling components due to the generated heat in the cooling stack. However, if this stack can generate less heat, then radiators, pumps, and coolant will not be required.

The standard for a modern vehicle requires it to start within 2 seconds at worst. A fuel cell starts well within 1 second. However, fuel cells, including hydrogen fuel cells, do not operate well at subfreezing temperatures. This is because fuel cells are basically a liquid interface device and need liquid-phase water to operate. Running the system under the conditions of a highway environment is possible, but the current cost is too great for commercialization.

HYBRIDIZATION

Hybridization is the optimization of two different power plants with two different characteristics and blending them together into a system that is better than either of the two components alone. In the case of transportation, hybridization involves a fuel cell and a battery or a battery and an internal combustion engine. The stacks, its fuel cell system, and the battery are completely different devices with different characteristics, so a tradeoff needs to occur. Fortunately, energy can be recaptured, transient capability can be improved, and the fuel cell can be put into its most efficient operating mode to literally use a battery to carry out all the operating modes that are detrimental to the fuel cell and to only allow the fuel cell to operate in a benign environment. The two devices working together can also help improve the start time. With hybridization, the size of the stack can be reduced, in turn reducing overall costs.

Of course, the negative tradeoff is that an extra component is added. The battery has its costs with benefits that need to be quantified. However, the addition of a hybrid battery up to the peak of 60 percent of a 60/40 battery/fuel cell stack yields a 50 percent improvement in fuel efficiency. While the battery has no ability to provide independent primary energy, hybridization is still an amazing new technology, especially since speed, start-up launch, and performance improve for the customer.

INSTITUTIONAL STRUCTURE

More is gained by going out to the interfaces and solving problems as a system rather than trying to solve everything one component at a time—for example, at the membrane or at the fuel cell stack. This is also true for corporate structures. DaimlerChrysler and Ford have a partnership in the fuel cell area. Until fairly recently, it was a rather complex corporate partnership with other companies. For example, Ballard had responsibility for the stack, Xcellsis for the fuel cell system, Ecostar for the power electronics and drive train, and Ballard Automotive for trying to integrate it to some extent and sell it as a unit.

This corporate structure added corporate and cultural interfaces into the system's engineering tasks. In the recent reorganization, all of the system interfaces are now contained inside one corporation, which solved a lot of issues in terms of development interfaces and a lot of business issues as well. The message here is that interfaces need to be addressed not just in terms of processes but also in terms of the vital institutional structure that surrounds it.

HYDROGEN STORAGE

Practical use of hydrogen in vehicles may never happen until there is a better method to store hydrogen, especially since onboard reforming of hydrogen at a reasonable cost may not be a possibility. The U.S. Department of Energy has worked with the auto industry and has ranked the options for hydrogen storage. The best candidate so far is compressed gas at pressures of about 700 bar. While not an ideal solution, it is probably marketable.

The use of hydrogen requires additional infrastructure for production and transportation. One method is to use electrical energy to produce hydrogen, but power grids are very inefficient. Another is the use of a natural gas pipeline, which is also wasteful since it involves the liquefying and re-evaporation of gases. The issues that need to be addressed in terms of fueling are the required number of fueling stations, the amount of money it will cost, how the oil industry will react, and how the auto industry will manage this emerging technology.

THREE-WAY INTERFACE

Renewable energy is the ultimate goal. Over the very long term, there must be a sustainable energy system that does not require a depletable resource. Unfortunately, at this time there are no systems that are economically competitive and renewable. The Department of Energy supports this goal, but it will be universities or the private sector that reach it.

Fuel flexibility and new infrastructures in fuels are three-way issues between the auto industry, the energy providers, and the federal government. Dealing with this three-way interface will be a problem, especially since each institution is used to having its own way. Fortunately, there is an institution like the California Fuel Cell Partnership, which is not just localized to California but is the world's fuel cell partnership. It is the only institution in the world that involves major public sector players, the major energy providers, and the major auto companies of the world. This is the only forum where they can all interact and struggle with the fuel cell problem.

The partnership is actually working. An initial report on the infrastructure problem has been issued recently. Although it really raises more questions than it provides answers, it puts forth a common language for all involved parties to use to talk about the problem.

Recently, there has been an announcement that Toyota has created a new organization inside the company. A fuel cell technology center has been created that includes vehicle development, fuel cell system development, fuel cell stack development, and manufacturing. Toyota also has run into this interface problem in its development efforts and consequently has created an organizational answer to it that allows the company to solve this giant interface issue across all these areas. DaimlerChrysler and Ford have both reached the same conclusion. There needs to be more openness toward creating new institutions to solve fuel cell problems. Fuel cell development is a major systems engineering issue, and the problems involved will not be solved simply by applying current processes. Fuel cell issues need to be solved now even if the solutions are not needed immediately. Waiting until fuel cells are needed may be too late to solve the problem.

4

Interface Challenges and Opportunities in Energy and Transportation

James R. Katzer,
ExxonMobil

INTRODUCTION

Future challenges in energy and transportation depend on the projected costs for energy options. These options are dependent on world energy demand, renewable energy availability (particularly photovoltaic refining), hydrogen availability as an energy resource, and light-duty vehicle power train changes.

In general, as the world's economies grow, the world energy demand also will grow. The magnitude of this growth is expected to be large—from about 200 million barrels of oil equivalent a day to over 300 million barrels per day in 2020. Given the expected rates of economic growth, energy demand is expected to continue to grow through 2050.

Fossil energy sources will continue to dominate through at least the first half of the 21st century because they are plentiful, readily available, and fundamentally inexpensive. Currently, 75 million barrels of oil are consumed every day. By 2020 it is anticipated that 112 million to 114 million barrels of oil will be consumed daily. Availability of oil will not be a problem outside of geopolitical issues that may appear. In fact, in the last 20 years, the proven reserves of crude oil have increased by over 50 percent from a little over 600 billion barrels to about 1 trillion barrels of oil.

RENEWABLE ENERGY

Renewable energy technologies are expected to grow significantly. However, over the next 20 years even with this rapid growth, renewable wind and solar energy will contribute only about 1 million barrels of oil equivalent per day of energy out of 300 million in 2020. Beyond 2020, growth of renewable energy technologies could be very rapid, but it is projected that this technology will produce less than 3 percent of the total energy demand.

The main forms of renewable energy have a significant hurtle in terms of cost. While there appears to be a consensus for increased use of renewables for electricity generation—the largest user of primary energy at present—the cost of renewables must be addressed if this is to take place.

Significant progress has been made in reducing the cost of renewables, but they are still limited in applicability by cost. Costs can be expected to decrease in the future, but the rate and extent will be critical to the rate of growth of renewables.

Photovoltaics and New Materials

Although the costs of generating electricity using most sources of renewable energy are still very high, they have been decreasing over the past 20 years and should continue to decrease. Wind is currently cost competitive for electricity generation, and while photovoltaic solar technology is improving, it is still an order of magnitude too expensive to be competitive.

Module efficiency is a key parameter for photovoltaics because cost per watt generally decreases as module efficiency increases. Module material costs and manufacturing costs per unit are fairly fixed. Thus, power costs decrease as module efficiency increases.

Advanced materials such as polymers that can reduce module and manufacturing costs may help photovoltaic technology become more competitive. Crystalline silicon module costs have decreased in the past 20 years but remain too high to be broadly competitive. Even the use of glass is too expensive compared to other low-cost power-generating technologies. Improvements in manufacturing processes also are needed to reduce costs and increase reliability. Power electronics must be improved, but there is additional cost related to energy storage. Major reductions in every component associated with the system are necessary to make photovoltaics a viable method of energy capture and storage. This will require research at the interfaces between the various chemical sciences as well as at the interfaces with theoretical physics, solid-state physics, materials science, electrical engineering, and manufacturing technologies.

Fuels for Transportation

The main source of energy for transportation is crude oil. In 1980, world oil demand was 65 million barrels per day. For the year 2000, actual world oil demand was 75 million barrels per day, essentially in line with projections made by Exxon in 1980. Proven reserves increased significantly between 1980 and 2000, due mainly to the impact of advanced technologies and increased exploration.

Crude oil refining is a key factor in keeping energy costs down using oil. Refining today is a sophisticated process that uses advanced technology to convert crude oil into a range of products, including transportation fuels, lubricants, petrochemicals, and lower-value so-called black products.

Depending on its configuration, the energy efficiency of a typical refinery is between 87 and 92 percent. Most of the energy in the crude is consumed in separations in the refinery; more than 90 percent of this energy is used in distillation.

Over 80 percent of the products from a refinery are transportation fuels and on-road fuels that are being driven to be more chemically specific through both regulation and the needs of vehicle power trains. Many other products are by nature chemical specific. Tightening of tailpipe emission standards is driving sulfur in gasoline and diesel fuel to ever lower levels. This is to allow after–treatment technologies that are focused on reducing nitrogen oxide and hydrocarbons to extremely low levels. Concerns over air toxins will also impact the chemical composition of gasoline in the future.

Automakers' search for more efficient power trains will consequently produce changes in the fuel that is needed to operate those power trains. This, too, will require chemical specificity, in the form of composition-based refining. The keys to molecular specificity in the refinery are accurate, chemically specific measurements; analysis; chemically specific processes; and real-time optimization of the integrated refinery. Today, the petroleum industry's ability to accomplish this is rapidly improving. For example, high-detail hydrocarbon analysis allows measurement of composition in great detail and structure-oriented lumping enables modeling of reactions with high precision. With these capabilities and with process models that now can predict product composition and properties from the crude to the end products, the foundation exists to do real-time optimization on the activities of the integrated refinery. Despite of improvements and new technologies, there are related needs that can be met by research in applied physics, materials science, chemical sciences, chemical engineering, and applied math.

THE REFINERY OF THE FUTURE

The refinery of the future will be more technology focused than today. It will make only high-value products, with one of those products being power. The refinery will be a clean, smart, high-value, energy-efficient installation—"clean" in that it will be environmentally benign, and "smart" in that it will be highly integrated, with operations managed around quantitative chemical reaction engineering models.

Chemical specificity will be achieved in the refinery of the future through the use of new catalysts. Catalysts are clearly a tool for creating and managing the desired chemical specificity. Shape-selective zeolite catalysts have been very effective in managing molecular composition based on size and shape. High-

> **Sidebar 4.1**
> **Online Analysis and Controls**
>
> A modern-day petroleum refinery is a complex chemical operation that involves numerous separations and chemical processing steps. Today virtually all the chemical analysis equipment found in the research laboratory is also used in the refinery or an online basis is often coupled to a control circuit to monitor product quality and make the necessary immediate adjustment in process conditions required to meet product specifications. While the online gas chromatograph is the most widely used instrument, infrared spectrometers, mass spectrometers, pH indicators, new infrared spectrometers with chemometric capability and moisture analysis based in solid-state conductors are not found in every refinery in the country. Until the 1970s, samples of most process streams in the refinery were taken at periodic intervals during the day and adjustments were made after the research was received from the refinery's analytical lab. This process was followed by the installation of online analysis equipment that sounded alarms, and the equipment operators took appropriate action. Today most operations are on computer control and the information received from online analytical equipment is processed almost continuously and controls make the required changes. An alarm may still sound and the equipment operator still responds, but usually the problem has already been corrected.
>
> In-line analysis and control systems that prevent the rerunning of a nonspecification product are safer and certainly better way to run a complex operation. While samples are still taken and analyzed in the laboratory, the research is mostly used for calibration purposes and to check online analysis.

activity, site-specific, supported catalysts also play a critical role in refining molecular management. However, much fundamental science needs to be developed to allow catalysts to become more effective in operations. Today, most multistep reactions are carried out a single step at a time followed by separations, resulting in high-energy consumption and byproduct formation. Better approaches to chemical reactions are required, including catalysts with the appropriate balance of activity in multifunctional capabilities that lead to successful execution of multi-step reactions either with a single catalyst or in a single reactor.

For these capabilities it may be instructive to learn from nature, which performs mulitstep reactions very efficiently. Enzymes and biological systems under a wide range of conditions carry out a very sophisticated array of reactions with high selectivity and essentially no undesired byproducts. Biocatalysts and pro-

> **Sidebar 4.4**
> **Zeolite Catalysis**
>
> Zeolites are microporous inorganic crystalline solids with ion exchange properties. Zeolites may be modified to enhance their catalytic activity. In addition, metals such as platinum can be added to introduce new catalytic functionality. Because hydrocarbon molecules having desirable fuel properties can be readily absorbed into zeolites, they have found great utility as catalysts for fuel processing. Zeolite catalysts have revolutionized the refining of petroleum and the manufacture of petrochemicals. The leading application of zeolites is for catalytic cracking of high-boiling petroleum fractions into gasoline and diesel fuel, thereby increasing the gasoline yield per barrel of crude by over 50 percent versus the prior technology using amorphous catalysts. Also, by using other synthesized "shape-selective" zeolites, a wide variety of petrochemical processes with improved efficiency and environmental performance have been introduced. The discovery and characterization of new zeolites continue. Recently a new meso porous zeolite-like family of materials (MCM-41) that can be made with uniform pores that can range from 15 to over 100 Å was discovered. This has caught the interest of the scientific community because it opens up potentially catalytically useful materials having highly controllable pore sizes and surface functionality for new catalytic processes.

cesses could be adapted for use in a refinery, where bioproduced fuels could also be a feed- or blend stock. More futuristic ideas include adaptive catalysts that respond to changes in feedstock, impurities, and the product needs of a refinery and that can be driven by some outside signal, such as microwaves and electronic or magnetic radiation. For all of these developments, interfaces are needed between the chemical sciences and materials science, physics, and the biological sciences.

Membranes play a more important role in the refinery today and will do so to an even greater extent in the future. They promise energy efficiency and chemical specificity, two of the most critical issues for the refinery of the future. Approximately 10 percent of the energy in crude oil is used or consumed in the refining process. Membranes have the potential to markedly reduce this energy consumption, increase refinery efficiency, and decrease the emissions associated with refining. For example, membrane separations could reduce capital and operations costs for the separation of propane and propylene by 50 to 70 percent. Savings due to reduced energy costs could be between $0.6 and $0.9 per pound, which is roughly half the cost of the propylene.

Combining membranes with catalytic reactions offers new opportunities for increased specificity and lower selectivity reactions as well as reduced costs. Such a combination could be applied to air separation, gasification, and large downstream separations. The challenges are to find membranes with high fluxes that are highly selective and durable under the conditions in which they operate in the refinery, field, or reservoir. These are significant challenges with possible significant results, especially if they affect hydrogen sulfide and carbon dioxide separations, which consume significant amounts of energy. Membrane development requires work across many chemical science interfaces, with much to learn from the biosciences. The future of research is at critical interfaces among the manufacturing sciences, device development sciences, and engineering.

There are still many needs required for improvements to refineries. One is a wide array of sensors, which will help to provide the foundation for real-time optimization of the integrated refinery and help to manage every point in the refinery. To help meet these needs, a key interface must be developed between chemical engineering and applied mathematics.

IMPROVING VEHICLE FUEL EFFICIENCY

Changes in automotive light-duty vehicle power trains are being driven by the need to increase energy efficiency and reduce carbon dioxide emissions. Conventional gasoline today is highly efficient and has very good emissions performance. To increase energy efficiency, additional technologies are being investigated very aggressively by the auto industry. Technologies also need to be developed to meet increasingly tough emissions standards that require reductions in nitrogen oxide from gas-powered engines. Nitrogen oxide reductions have been made continually over the past 30 years. The newest regulations, federal Tier 2 standards that go into effect in 2004, effectively target other cold start pollutants such as nonmethane organic gas emissions without affecting nitrogen oxide control. Extremely high levels of simultaneous reduction of NO_x, carbon monoxide, and hydrocarbons have been possible because of the stoiciometric nature of engine-out exhaust of the internal combustion gasoline engine and the development of three-way catalyst technology. To meet current or future emissions reduction legislation, there is a push toward more sophisticated models of engine operation with predictive capabilities and real-time optimization.

Meeting even higher levels of emissions controls will require further technological advances. Catalysts with lower thermal mass and light-off temperature that can be coupled to the exhaust manifold will be required. Catalysts are needed to very efficiently adsorb emissions during cold-start operations. Catalysts are also need that can convert these adsorbed materials when they desorb. Sophisticated models of energy operation, including predicting emissions when going from transient cold-start to full-throttle operation, coupled with a catalyst after treatment operating mode that can deliver precise individual-cylinder air/fuel con-

> **Sidebar 4.2**
> **Importance of Environmental Process Knowledge to Energy and Transportation Systems**
>
> Much of the process innovation since 1970 has been driven by the need to reduce the undesirable environmental impacts of emissions (both process and exhaust) from energy and transportation system components.
>
> Emissions from energy and transportation systems are known to affect air pollution/air quality issues on a local to global scale, including photochemical smog, urban and regional haze, fine aerosol impacts on lung function, acid rain and acid deposition, greenhouse gases and aerosol impacts on climate, and stratospheric ozone depletion. Point and mobile source emissions of gases, aerosols, liquid wastes, and solid wastes also impact surface and groundwater quality and ecosystem vitality on local to regional scales. Better understanding of the chemistry of air, water, and ecosystem pollution, including the kinetics of pollutant transformation and the dynamics of pollutant transport processes, has greatly clarified how emissions from energy and transportation systems adversely impact environmental components. This allows a systematic analysis to prioritize emissions reduction requirements and a scientifically sound capability to decide how much specific emissions need to be reduced to achieve environmental and health goals. Process knowledge of the full range of environmental issues also allows identification of when a current or proposed emissions reduction strategy may exacerbate another environmental problem.
>
> The same research tools used to achieve a far more fundamental understanding of environmental processes, including: highly sensitive and specific chemical sensors with short response times, integrated multispecies and physical property sensor suites for field deployment, and comprehensive fluid dynamic/chemical kinetic computer models of environmental systems, can also be used to more thoroughly understand the chemistry and physics of energy and transportation system components. Subsequently, the tools needed for progress in both fields are often interchangeable, benefiting both.

trol must be developed. Technological advances such as these should allow for further emissions reduction by a factor of 2 to 10.

Lean-Burn Technologies

Lean-burn engine technologies such as gasoline direct injection and diesel engines can provide significant improvements in fuel efficiency and performance.

In these engines, fuel is injected directly into the combustion chamber, forming fuel-rich and fuel-lean regions. The overall combustion mixture is typically lean. These engines operated over a wide range of in-cylinder combustion regimes.

Diesel engines, however, are well known for soot particle production, and both diesel and gasoline direct injection engines produce large amounts of nitrogen oxide. Furthermore, when operated lean, these engines present enormous challenges in nitrogen oxide removal because the exhaust is strongly oxidizing, and three-way catalysts developed for conventional gasoline engines are not applicable. To use these lean-burn engine technologies, entirely new after-treatment technologies will be required.

Resolution of these challenges will require an approach that integrates key chemical sciences disciplines. Included are fuel science and combustion chemistry, thermodynamics, and kinetics to understand the combustion process and the formation of particulates and nitrogen oxide. Computational fluid dynamics is needed throughout the entire system. This must be coupled with combustion kinetics to predict the course of the combustion process and to identify strategies for controlling nitrogen oxide and particulate formation. Engine design and control skills are required to convert theoretical understanding into real-world performance. Catalysts and reactor design and development are necessary for after-treatment systems to trap or reduce nitrogen oxide and particulates to very low levels. Finally, all of these needs must be predictive, not in the steady state but in the highly transient operation that the system continuously performs.

Fuel Cells

Current interest in fuel cell vehicles is focused on increased fuel efficiency, reduced or zero emissions, and enhanced performance to meet consumer needs and expectations. Fuel cell vehicles offer great opportunities and challenges. Current proton exchange membrane fuel cells require hydrogen to produce electricity. This raises a myriad of challenges regarding the source, production, and storage of hydrogen.

The classical model of fuel supply would favor off-board hydrogen generation and refueling at stations, much the way gasoline is managed today. However, there is no hydrogen fueling infrastructure today, and infrastructure development is costly. Onboard hydrogen generation from available fuels such as gasoline avoids infrastructure challenges but has vehicle technology and cost challenges.

The experience of the process industry is that, as the scale of a process is increased, the cost of the product per unit is reduced. For hydrogen generation at very large scales, production from natural gas is possible for slightly more than the cost of gasoline. However, as plant size decreases to the range of a station servicing an average of 150 fuel cell vehicles, the cost of hydrogen fuel is higher because of the capital cost associated with the facility and the higher cost of

> **Sidebar 4.3**
> **Solar Photoconversion**
>
> Photoconversion is the process whereby the energy of solar photons is converted directly into electricity or into stored chemical-free energy in the form of liquid or gaseous fuels. The former process is termed *photovoltaics* and can be achieved using photoactive inorganic and organic semiconductor structures (such as p-n junctions and other types of interfaces) or in photoelectrochemical cells based on interfaces between semiconductors and chemical redox systems. The latter process is also based on photoelectrochemical structures and produces fuels, such as hydrogen, hydrocarbons, and alcohols, from simple substrates like water and carbon dioxide. Photoconversion is a renewable energy technology that offers the possibility of providing an unlimited energy resource that is also free from carbon emissions. The challenge for chemistry is to make photoconversion sufficiently efficient and inexpensive, and for the case of photovoltaic conversion to also provide practical and cost-effective electric power storage.

hydrogen production. The challenge to break this paradigm requires major development and breakthrough.

The method of off-board hydrogen generation greatly affects its cost. Hydrogen produced in large-scale refineries is estimated to cost $0.85 per gallon, while small-scale hydrogen generation at a vehicle refueling station is expected to cost $6 to $8 per gallon.[1] These added costs result mainly from distribution, which involves compressing hydrogen to high pressure of 6,000 to 7,000 pounds and storing it in cylinders at the station. The associated capital cost is more than double the cost of fuel delivered to the vehicle.

Energy storage density is a major issue because the low energy density of hydrogen makes it expensive to transport and store. Automobile companies and energy companies have the same storage problem. Breakthroughs in understanding system integration modeling are needed. A major breakthrough in hydrogen storage will be required through innovation and new technologies from all the chemical, materials, and engineering sciences, solid-state physics, and all the interfaces associated with these fields.

The challenge for onboard generation of hydrogen is no less monumental. To be competitive with the internal combustion engine, the fuel cell engine must cost approximately $3,000. This means that for a 100 kW power source the onboard

---

[1] Donald Hubert, President, Shell Hydrogen, presentation to the National Hydrogen Association.

fuel processor needs to cost approximately $15 per kilowatt. This is less than one-tenth the cost of a large-scale, highly efficient, steady-state methane reformer for hydrogen production. In addition, the onboard fuel processor must operate in a highly transient mode.

Innovation will be essential in almost every aspect of the fuel processor to achieve this goal. Innovation in materials, including both catalysts and membranes, separations and hydrogen storage, systems integration, and manufacturing will all be necessary.

A potential solution to these challenges may be the generation of hydrogen from renewable sources, such as sunlight. However, until photovoltaic electricity costs decrease, there will be no solar hydrogen generation at a competitive cost. Again, innovation in materials and manufacturing will be necessary.

CONCLUSIONS

For energy and transportation in the next 20 years the chemical sciences are positioned to be the key sciences involved in resolving the issues and moving society forward. Fundamental innovations will be required, particularly in the area of materials chemistry. The challenges are becoming increasingly complex and interdisciplinary, which will require working across all of these interfaces to have progress.

The concern will not be a lack of innovations, since an innovation does not denote a successful application. The challenges for the chemical sciences community will be to sift through those innovations, repeatedly refocus them, drive them through development as a multidisciplinary team, and diffuse them out into the entire economy. This is the biggest challenge that the community will have in the future.

5

R&D Challenges in the Chemical Sciences to Enable Widespread Utilization of Renewable Energy

Nathan S. Lewis,
California Institute of Technology

When looking at the present utilization of primary power and options for future sources of energy, a number of questions can be envisioned. Where do we presently get our power from and what are the costs? What is the role for the chemical sciences in renewable energy technology? To answer these questions, it is useful to review the present primary power mix—how much energy is consumed, from what sources, future constraints imposed by sustainability, and the theoretical and practical energy potential of various renewables. Once these issues have been addressed, it is possible to identify the challenges for the chemical sciences to economically exploit renewables on a scale commensurate with our energy needs.

The mean global energy consumption rate in 1998 amounted to 12.8 TW (383 quad/year),[1] of which only 10 percent was used for electricity (Figure 5.1). Oil, gas, and coal constitute almost 80 percent of this total energy consumption. The mean U.S. energy consumption rate in 1998 was 3.3 TW (99 quad/year), 15 percent of which was used to generate electricity.

At present, the use of renewables for energy accounts for only a small percentage of total energy requirements. Of the 12.8 TW of global power consumed in 1998, power from biomass accounted for only 1.21 TW, while the largest renewable source consumed—hydroelectric—still accounted for only 0.3 TW.

One possible explanation for the limited use of renewables may be cost. With one exception, the use of fossil fuels for electricity generation in the United States is presently by far the least expensive option. The cost per kilowatt-hour for

---

[1] 1 quad = 1015 Btu = 1.055×10^{18} J; 1 quad/year = 0.0334 TW.

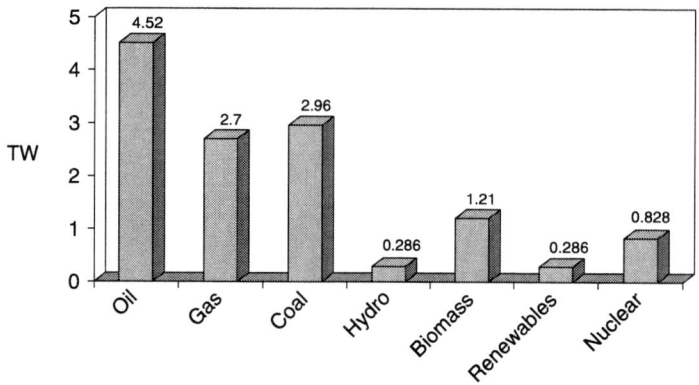

FIGURE 5.1 Mean global energy consumption, 1998.

electricity derived from coal is 2.1 cents; for natural gas is 3.6 cents; and for oil, 3.9 cents. This compares with a cost per kilowatt-hour for wind of 3 to 5 cents, and for solar approximately 22 cents. However, it must be noted that electricity derived from nuclear power is quite competitive with fossil fuels, with a cost of 2.3 cents per kilowatt-hour.[2,3]

Proven reserves of oil, natural gas, and coal tend to underscore that fossil fuels will remain an abundant inexpensive resource base for the foreseeable future. Proven oil reserves are expected to last at least 40 years, natural gas should last at least 70 years, and coal reserves are adequate for 200 years. By adding reserves that are likely to be found, the oil supply would then last between 50 and 100 years, natural gas 90 to 275 years, and coal at least 2,000 years.

In light of this supply, it is reasonable to conclude that renewables will not play a large role in primary power generation unless one of two things happens. Either technology breakthroughs on these renewable sources of energy reduce the cost of these sources significantly below what they are today, or there are unpriced externalities introduced that significantly increase the cost of fossil fuels. For example, environmental concerns may result in the introduction of carbon taxes or subsidies for carbon-neutral technology.

Looking to the future, it may be possible to make some predictions regarding primary power demands and the expected environmental impact over the next 50

---

[2]These costs represent the costs of capital equipment and fuel but not waste disposal or environmental remediation.

[3]Electricity, however, is high-value energy, and the cost per Joule of energy produced by consuming fossil fuels to make heat is approximately a factor of 5 to 10 times less expensive.

to 100 years based on demographics. The global population is expected to rise to 10 billion–11 billion people by 2050, and GDP growth is expected to rise an average of 1.6 percent a year, which is its historical average. Balanced against these factors, energy consumption per unit of gross domestic product (GDP) is expected to decline 1 percent a year, mainly as a result of increases in energy efficiency in the industrialized world.

Based on these projections, total primary power consumption is expected to rise from 12 TW in 1990 up to 28 TW by 2050. The overriding question is from where this 28 TW of power will be derived.

The above projection, as previously stated, depends on increased efficiency in energy usage. Sustaining the historical trend of carbon intensity in the energy mix implies that by 2050 the energy economy will actually be more efficient than one run entirely on natural gas. This can only be accomplished if there are significant contributions from carbon-neutral power. Based on projected carbon dioxide emissions from fossil fuel usage, it is estimated that even with increased use of fossil fuels over the coming century, these fuels will not be enough to meet the expected 28 TW of power demanded by 2050. There will be an estimated 10 TW shortfall that must be derived from carbon-neutral sources in order to meet the primary power needs of the planet.

Obviously, this is not an insignificant amount of power. Ten terawatts was the entire global power production from all sources in 1990. In addition, if atmospheric carbon dioxide emissions are to be stabilized, there will be an even greater need for renewable energy. For example, if man-made releases of carbon dioxide are to be stabilized at 550 ppm—twice the preindustrial level of carbon dioxide in the atmosphere—20 to 30 TW of carbon-neutral primary power would be needed.[4]

This is a daunting amount of power. Without economic policy incentives, the needed technology to meet these demands will probably not be in place soon enough to meet this demand by 2050. In fact, meeting the goal of commercial carbon-neutral power capable of producing 10 to 20 TW by the mid-21st century could require efforts comparable to the Manhattan Project or the Apollo space program.

Where such large amounts of power can be derived from must be examined. If this power is to be carbon neutral, the technologies needed must also be examined. To develop these technologies, the challenges for the chemical sciences must be identified.

The five most common renewable sources of primary power are hydroelectric, geothermal, wind, biomass, and solar. Hydroelectric power is considered by many to be a model energy source. It is clean, relatively benign environmentally, nonpolluting, and relatively inexpensive. However, the global theoretical potential of the hydrology of all the world's precipitation and all of the energy

---

[4]Hoffert, M. I. et al., Nature, 395:881. 1998.

flows of the waters on Earth provide only 4.6 TW, far below the goal of 20 TW of power. The technically feasible amount of hydroelectric power is far less, only about 0.7 TW. The installed capacity of hydroelectric power is already 0.5 TW. Therefore, there is not much additional power available for exploitation from this source.

Power derived from wind presents significant onshore potential. Two large geographical areas—the Great Plains of the United States and the region from Inner Mongolia to northwest China—present significant expanses of land suitable to utilize wind. If 6 percent of the dedicated land in the Great Plains were used for wind farms, approximately 0.5 TW of power could potentially be obtained. As with hydroelectric, this is a significant amount of power but far short of the projected 20 TW of additional power that will be required.

Globally, it is theoretically possible to obtain 50 TW of power from wind stations on land, but with practical land usage—about 4 percent of all the land that has enough wind to make power generation economically feasible—the potential amount of power that can be derived is approximately 2 TW. The offshore potential for wind power generation is larger, but there is the significant requirement of being close to an electrical grid to make it practical.

Distribution is a key concern with wind generation of power. If, for example, the Great Plains were used to generate large amounts of power for the United States, this power would not be consumed locally, and there are constraints to the electrical grid. An efficient method of power storage would have to be found. Presently the methods available for power storage carry too high a penalty in terms of energy loss to make them economically feasible for wind as a significant source of energy.

Biomass as a source of a large percentage of the world's power has significant obstacles. Biomass requires large areas because the process is very inefficient. Only 3 percent of the total sunlight that is incident per unit area on a plant is actually stored in free energy by photosynthesis. This is sufficient for biological needs but is difficult to exploit as a source of primary power. To meet the goal of 20 TW of additional power, biomass would require 4×10^{13} m^2 of land, and the total landmass of the earth is 1.3×10^{14}. Clearly this is not a viable option.

It is possible to look at the situation with biomass from a different perspective. The amount of land with crop production potential in 1990 was 2.45×10^{13} m^2. In order to support 9 billion people in 2050, 0.416×10^{13} m^2 of additional land will be required for crop production. The remaining land available for biomass energy would then be 1.28×10^{13} m^2. This would result in a projected total of 7 to 10 TW of power.

This would be a massive undertaking, requiring that almost all of the crop production potential on the planet be utilized. In addition, there are significant obstacles, not the least of which is the issue of water resources. Also, cellulose derived from biomass must be readily converted to a liquid fuel—preferably ethanol—presenting a challenge for the chemical sciences.

Nuclear power, which will be discussed in greater detail in other presentations in this workshop, would require 10,000 new 1 gigawatt nuclear power plants in order to provide 10 terawatts of additional energy. Building one of these power plants every 2 miles along the California coast would provide only 300 of the needed 10,000 power plants. Clearly, it will not be practical for nuclear power to be the sole source of this additional power.

Solar power is still another option for noncarbon primary power. Theoretically there is 1.2×10^5 TW of solar energy potential. However, if solar cells are assumed to be 10 percent efficient, realistic land estimates lead to a practical value of 600 TW of available incident solar power, leading to 60 TW of generated power with 10 percent conversion efficiency.

To generate 20 TW of power using solar cells with 10 percent efficiency requires approximately 0.16 percent of the world's landmass, including 8.8 percent of the landmass of the United States. To generate 12 TW of primary power by this method would require 0.1 percent of the Earth's landmass, including 5.5 percent of the United States.

These numbers are still quite large, but compared to other methods for generating noncarbon primary power, solar power appears to be the most compelling method. To achieve anything approaching these numbers using solar energy requires one of three approaches. A low-efficiency low-cost method is through photosynthesis. Alternatively, a highly efficient process yet high-cost method uses photovoltaics. A third method utilizes semiconductor liquid junctions and photocatalysts with the ultimate goal of using sunlight to split water into hydrogen and oxygen or to make electricity. Both the cost and efficiency of this process are moderate.

Production capacity for solar electricity is currently limited to about 100 MW per year, even though it is a subsidized industry. This industry is growing rapidly—on the order of 30 percent per year—although it must be noted that this is from a small base. Solar electricity currently makes up 0.1 percent of total electricity production.

The rate of progress for a variety of different technologies shows increasing photovoltaic efficiency, yet most of these technologies have to contend with physical limitations. Silicon or crystalline semiconductors have very high conversion efficiencies, yet these crystals are very costly to make because the grain sizes must be large. When smaller grain sizes are used, these semiconductors are cheaper to make, but they have much shorter lifetimes. Single-crystal silicon can be replaced with a less expensive organic material, but these organic films currently also have short lifetimes and therefore produce devices with low efficiency.

Based on these physical limitations it appears unlikely that, with normal market forces and normal research advances, these technologies will provide the amount of power needed economically by 2050. What is likely to be needed is the implementation of a new solar technology that initially offers less performance but at a far lower cost, and one that eventually overtakes present technologies as it improves.

In terms of technologies for photovoltaics, a grand challenge would be to devise alternatives to massive single crystals, which clearly cannot be produced economically. One way to achieve this goal would be to passivate grain boundaries in order to get polycrystalline samples to act as a large single crystal. Electron transfer agents could be used to link together grain boundaries, thereby allowing electrons to move from grain to grain without inducing recombination at the grain boundary edges.

One particular method for achieving this goal that is currently being investigated utilizes titanium dioxide, an inexpensive pigment found in white paint.[5] The TiO_2 is coated on a glass slide and sensitized with a dye to absorb sunlight. Electricity can be generated with 5 to 10 percent efficiency. While there are still questions about the long-term stability of this material, it does represent a new and inexpensive approach to photovoltaics different than technologies presently used. Other approaches with different light absorbers, such as interpenetrating polymer networks, nanocrystals, and other inexpensive approaches to solar energy conversion, should be explored as well.

Photoelectrolysis is a technology that converts light into both electrical and chemical energy. Solid $SrTiO_3$ is used in a photochemical cell where it absorbs sunlight and effectively splits water with high quantum yield. Electrolysis of water using this process can be sustained almost indefinitely. However, the band gap for $SrTiO_3$ (3.4 eV), is in the ultraviolet range, and materials with a lower band gap either are not stable in water, cannot sustain the electrolysis of water, and/or cannot absorb sunlight efficiently. Catalysts are also needed to effectively convert the photogenerated charge into chemical fuels.

SUMMARY

To meet the increased demands for primary power in the 21st century, normal economic driving forces appear to indicate that the demand for 28 TW of power without unacceptable environmental consequences could result from a combination of wind, solar, biomass, and nuclear power. However, sources such as wind and solar require new technologies to effectively store and transport power with little loss.

Another important consideration with solar power as a source of energy is that it inherently provides electrical power. Only about 10 percent of energy consumption is presently in the form of electricity, whereas the other 90 percent is used for heating, transportation, and industry. Even if electricity were used to meet part of these needs, it would not be used to meet all of the remaining 90 per-

---

[5]B. O'Regan and M. Grätzel, Nature, 353:737. 1991.

cent of power consumption unless direct photochemical or efficient electrochemical methods for energy storage or fuel generation were developed.

To make fuels that are storable and transportable, there are two primary chemical transformations to consider. One is the conversion of carbon dioxide to methanol, and the other is splitting water. Methanol could be used in a fuel cell where it is converted to carbon dioxide. This must be converted back to methanol to close the carbon loop. Alternatively, if hydrogen is used as a replacement fuel for carbon-based fuel, a hydrogen fuel cell that utilizes the product of solar or electrical water splitting would be available for transportation.

Whichever of these alternatives is ultimately adopted, the need for additional primary energy sources is apparent. In addition, the case can be made for significant carbon-neutral energy systems in the future. These technologies present significant challenges for the chemical sciences. For solar power, inexpensive conversion systems must be developed that include effective energy storage. Advances in the chemical sciences will also be needed to provide the new chemistry required to support an evolving mix of fuels for primary and secondary energy.

6

Nano- and Microscale Approaches to Energy Storage and Corrosion

Henry S. White,
University of Utah

Investigations of electrochemical phenomena at chemical structures that are measured and defined on the nanometer-length scale constitute a rapidly emerging frontier of electrochemical science. Measurement of single electron transfer events, electrochemical syntheses of patterned and well-defined nanostructures, and electrochemical characterization of nanometer-scale surface features have all occurred over the past decade. Technological applications of these scientific advances are on the horizon. An example of this is in the fabrication of new higher energy density batteries, which require new scientific strategies in the assembly and deployment of nanoscale structures to control macroscopic electrochemical phenomena.

The three-dimensional electrochemical cell is a hypothetical device that illustrates how some of the advances in microscale and nanoscale electrochemistry over the past two decades may be applied to its construction (Figure 6.1). The three-dimensional electrochemical cell is a conventional battery in the sense that it has a cathode and anode, but they are configured in an interpenetrating array with electrodes anywhere from micron dimensions if they are prepared using lithographic techniques down to the nanometer scale.

The advantage of the three-dimensional electrochemical cell over conventional two-dimensional batteries is that the additional dimension, termed L, can be increased indefinitely. Increasing L results in an increase in energy capacity without any loss of power density in the cell, because electrons travel the same distance between the anode and cathode regardless of the thickness of the cell. This cannot be accomplished with a conventional battery. Potential applications for this technology are Micro-Electro-Mechanical Systems devices and other microelectronic devices.

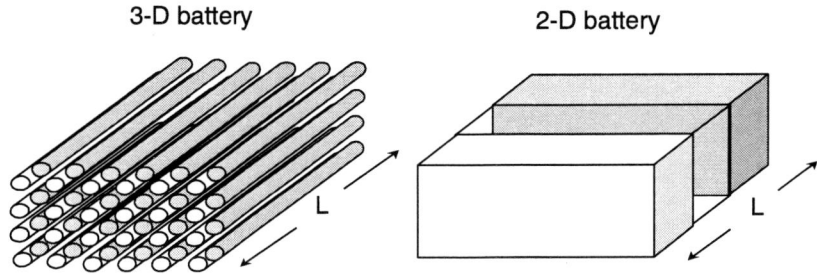

Increased capacity with L without losing power density per geometric footprint

FIGURE 6.1 The three-dimensional electrochemical cell.

Actual construction of such a device presents a number of technical challenges. When electrodes are synthesized at 10- to 100-nm diameters, with the anode and cathode separated by similar distances, problems in hard wiring and assembly are to be expected. In addition, there currently is no three-dimensional architecture for an electrochemical cell that would achieve uniform current density. Also, at the nanometer scale, noncontinuum effects, especially mass transport, become a concern. Other issues of concern include ensuring that there is enough territory for phase nucleation to occur and quantized charging when the electrode material approaches nanoscale dimensions.

While a simplified assembly may be pictured showing carbon nanotubes hardwired in an electronic circuit (Figure 6.2), it is difficult to envision how this could be accomplished using gold interconnects between carbon nanotubes. This approach is unlikely to be successful at least over the coming decade.

A different approach uses solid state chemistry to hard wire a three-dimensional battery-type structure. Solid gel chemistry has been used to produce a void-SiO_2 composite aerogel that has been electronically wired with a conformal nanoscopic web of RuO_2 to which 50- to 100-nm colloidal Au particles have been electrochemically attached (Figure 6.3). This configuration provides the contacts for collecting electricity out of the cell. Ruthenium oxide can act as one of the active electrode materials in the battery, and silicon dioxide (as opposed to a Nafion membrane) can act as the separating material. All that is needed to develop a truly functional three-dimensional battery is to put a second electrode material in place and make the electrical contact to the outside world. The result is a very high energy density device with no power losses because the anode and cathode are spaced together so closely. While this device has not yet been built, and the actual device may be several years off in the future, progress toward it has been made.

FIGURE 6.2 Hardwiring electrochemical nanocells.

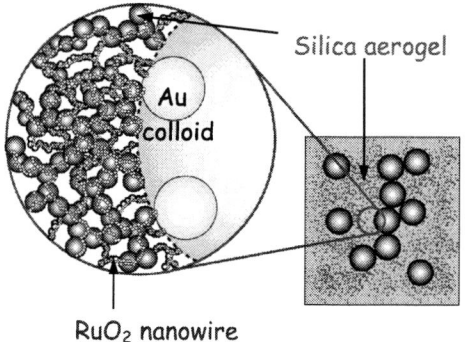

FIGURE 6.3 Schematic of a void-SiO_2 composite aerogel that has been electronically wired with a conformal nanoscopic web of RuO_2 to which 50- to 100-nm colloidal Au particles have been electrochemically attached. Ryan et al., *Nature*, 406: 169.

When making electrodes on the order of a few nanometers in diameter, as required for these cells, quantized double-layer charging must be addressed. With these small devices, electrons are not added continuously as a function of potential. Rather, electrons are added in discrete steps at different potentials as a consequence of the extremely small capacitance of the particle.

If this in fact is a quantized electrode, the resulting battery will be quantized and a discharge curve will have discrete steps. The result is that this type of device will have totally different operational parameters. Not only will these devices be based on different materials with different physical and chemical properties, but the output and function also would be inherently different.

Energy storage batteries are normally thought of as having a voltage determined by the free energy of the reactions at the anode and cathode. If the anode and cathode are very close together such that the electrical double layers begin to overlap, it is possible that the free energy of the cell will no longer be solely determined by the reactions at the anode and cathode but also by the electrostatics of the interaction between the electrodes.

Electrostatics is just one of the properties of materials that must be considered when approaching the nanoscale. The oxidation, kinetics, and general electrochemistry of materials have been observed to behave unpredictably at nanoscale dimensions. The fact that these properties are not well understood allows for many research opportunities as devices such as the three-dimensional batteries are made.

Because the three-dimensional battery has not yet been made, the question remains whether there is enough electrode area to have a nucleation event such that the anodic or cathodic reaction can occur.

THE CHEMISTRY OF CORROSION

In addition to energy storage, another broad challenge facing transportation research is corrosion. Many opportunities exist in materials chemistry related to the issue of corrosion, particularly when faced with challenges such as producing an economically viable automobile that achieves 80 miles per gallon fuel efficiency. One way to help meet this challenge would be to require a 40 percent reduction in the mass of the vehicle. One means to achieve this is to replace steel in automobiles with aluminum alloys.

Challenges like this will require a fundamental understanding of corrosion. Metallurgical issues such as the role of the preexisting distribution of elements in the alloys requires a detailed understanding of microstructure, which in turn is important in order to understand oxidative breakdown and the chemistry that causes the propagation of a stress corrosion crack, for example.

Work at Sandia National Laboratory has looked into the fundamentals of corrosion. Macroscopic results of experiments into the pitting of aluminum wire when exposed to sodium chloride solution indicate that the pitting potential is not a thermodynamic value but rather the potential associated with the kinetics of oxide breakdown. As a result, as a device becomes increasingly small, the probability of oxide breakdown will likewise decrease. At nanoscale a device made of this material would be highly unlikely to undergo oxide breakdown, and such a device would be expected to exhibit stability for long periods of time.

Characterization tools must also be developed if a fundamental understanding of corrosion is to be achieved. For example, observation of a titanium surface with an oxide film at the nanometer scale shows oxide grains on the surface. Conductivity atomic force microscopy measurements can be used to indicate defect-free TiO_2 by showing no current flow. This is visually indicated by a dark image.

Therefore, light spots on the image indicate defects in the oxide. With current analysis techniques, these defects can be found down to 10-nm resolution. However, fundamental structural information about these defect sites is still lacking because the analysis and characterization tools do not yet exist.

To advance a fundamental understanding of corrosion, researchers must move beyond empirical or phenomenological descriptions of corrosion mechanisms to a more molecular understanding. Characterization tools, including computational tools, are needed to achieve this goal. The overarching goal is to develop these corrosion mechanisms at a level of detail and sophistication similar to those found in chemistry and molecular biology.

7

Challenges for the Chemical Sciences in the 21st Century

Ralph P. Overend,
National Renewable Energy Laboratory

Predictions indicate that in the future energy usage will grow exponentially. In response to this growing need, it is also expected that new energy sources will be developed. The plot in Figure 7.1 depicts shares of the different energy forms. Basically, society has moved from a renewable, biomass, and wood economy in 1850 to a coal, oil, and gas economy by the late 1960s.

Key to the trend in increased future energy consumption is the anticipated growth of new energy sources: hydro, nuclear, intermittents that are a combination of solar and photovoltaic, and deep thermal sources. The most important message from Figure 7.1 is that somewhere between the years 2025 and 2050, there will be multiple sources of primary energy. Development of the various of energy sources is the first challenge to chemists and chemical engineers. There is no single solution. In addition to this challenge, the possibilities of carbon capture, geological sequestration, viable hydrogen systems, energy storage systems (these are crucial), and commercial biomass energy must all be considered.

Often, the use of biomass is considered a cycle in which the sun shines on a custom-grown plantation of biomass material for energy. This energy is then consumed in a convergence system, with residual carbon dioxide and water vapor escaping into the atmosphere only to be fixed back into the same plantation by photosynthesis. The truth is that the biomass renewable carbon system in the world's economy is very large. We eat bread. We wear cotton shirts. We live in wooden buildings. The fiber component in the United States alone represents between 5 to 7 percent of the total energy input to the U.S. economy. Crops and animals are not grown specifically for energy but instead are part of a sensible cycle using residues. Under those circumstances, biomass has an extremely large component contribution to this future energy supply.

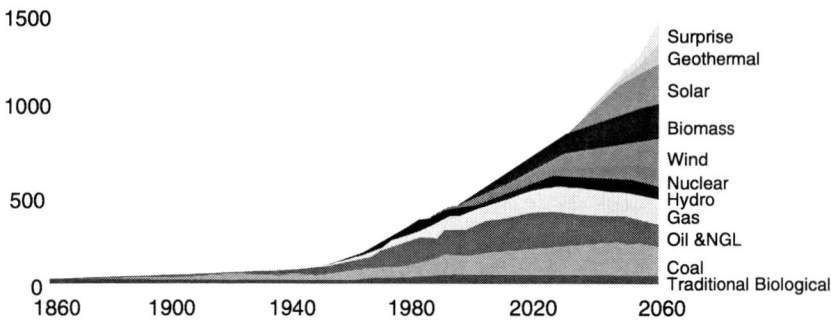

FIGURE 7.1 Shell's sustained growth scenario. The breakdown in growth of various energy forms is expressed in exajoules versus year. Source: *The Evolution of the World's Energy Systems*, 1995, Shell International Ltd, London.

Per capita power available in the United States has risen dramatically, from 200 to 300 watts per person in 1850, to close to 100 kilowatts per person today. Most is due to automobiles, which use between 60 and 130 kW. During the time period around World War I, the percentage of total energy from coal, oil, and gas was nearing 80 percent, concurrent with the mass use of the internal combustion engine during the war. Over the years a virtually unidimensional system has been created into which very diverse energy resources must be fit. That is a major challenge to scientists today.

The mechanical sciences have been the lubrication and the friction in a symbiosis between the petroleum industry and the transportation industry. On one side, refiners make different fuel, which have created the discipline of chemical engineering. On the other side, fuels conversion specialists have worked to make more efficient engines.

In the 1970s this unidimensional energy system was causing significant problems. Urban smog became the predominant feature of the West Coast, lead in gasoline was identified as a public health problem, and the first oil crisis occurred. However, these factors did not serve to hold public interest in fuel economy. Fuel economy became a popular topic because of clean air actions.

In the early 1990s the Partnership for a New Generation of Vehicles attempted to put together a vehicle that had high fuel economy, high emissions control, and very good environmental performance while safety and affordability were still maintained. These initiatives are beginning to have an impact.

In its energy predictions for 2030, the Finnish National Research Council shows that despite environmental issues the internal combustion engine has been actually saved from obsolescence by changes to the engine itself. These changes include exhaust gas recirculation, multipoint fuel injection, closed-loop air ratio control, direct injection spark ignition, and small diesel engines. Other improvements were made simultaneously: three-way catalytic converters, unleaded gasoline, detergent additives, oxygenated gasolines, and reformulated fuels. These scientific advances were made by multidisciplinary teams, including chemists and chemical engineers.

The development of catalysts in the engine system has evolved and has shown a significant impact. The 1970 engine had massive production of carbon monoxide, hydrocarbons, and nitrogen oxide. The way to control emissions was the three-way catalyst, but the overall energy efficiency was only 20 percent. Diesel engines reduced the levels of emissions and increased efficiency to 25 percent, but they have high particulate emissions. Oxycatalysts with a filtration system increases efficiency to 32 percent, but high nitrogen oxide emissions are still present.

This trend involves not just engines and the catalyst systems but also the fuels. When scientists talk about getting 80 miles per gallon, what is actually meant is 0.5 MJ per passenger-kilometer. Presently, the overall transportation system, not just autos, is actually running around 2 to 3 MJ per passenger-kilometer. There are many ways to improve efficiency besides working on the automobile.

The first way to improve efficiency is to use electricity. However, an electric system that can act like the gasoline or natural gas distribution system does not yet exist. Another problem with electricity is that batteries are still insufficient to run a vehicle for a day. Hydrogen is a solution but, like electricity, does not have the infrastructure required for mass production and use. Finally, many resources such as corn and forest residues can be converted through syngas into methanol. The question then becomes how the methanol is used: directly, through fuel cells, or as oxygenates in the traditional combustion system. The diversity of possibilities is part of the challenge. Many of the needed engineering and science tools exist, but there is a lack of focus.

Sidebar 7.1
Biomass Refining into Commodity Fuels

Biomass is the complex organic matter (carbohydrates, lignin, cellulose) produced by green plants and algae from water and carbon dioxide through biological photosynthesis. Biomass can be converted into various liquid and gaseous fuels (alcohols and hydrocarbons) either through biological processes (fermentation) or via thermal/catalytic processes (analogous to petroleum refining). Biomass represents a huge energy resource, and the challenge is to convert the myriad of biomass forms into commodity fuels with high selectivity, high efficiency, and low cost.

There are many predictions for the future. They range from the fuel cell electric drive vehicle by 2015 to the fall of pure petroleum fuels by 2030. A vision for the future based on the leaf may be considered. A leaf accumulates carbon dioxide, water, and photons; stores energy; and releases energy for its needs when it is wanted. Perhaps someday the personal automobile will, similar to the leaf, require no fuel other than sunlight. The skin would receive all the energy, be recyclable, be the energy storage medium, and use regenerable systems so that the solar energy received each day into the vehicle would be enough to provide for personal transportation needs. There are certainly some disadvantages to such a system, but these are merely challenges that must be overcome by chemists and chemical engineers.[1]

There are no single solutions to future energy needs, so chemists, chemical engineers, and other scientists must have a robust and diverse strategy to meet this need. The future will be varied with respect to energy. Energy supply and demand, and therefore infrastructure, may even be regionally different. Challenges lie in these many different production systems.

---

[1] One of the challenges is the amount of power possible from such a system. If a large car has an incident area of 7.5 m^2, this equates to about an available solar power of 10 kW. The current power requirement for an automobile to cruise at about 60 mph is approximately 40 HP. This is about 50 kW, which is where the design target is for most fuel cell systems. During this drive, even at 100% efficiency, the solar flux is insufficient for driving. In this case, an additional charging source will be necessary.

8

A Renaissance for Nuclear Power?

Patricia A. Baisden,
Lawrence Livermore National Laboratory

There are many reasons why nuclear power might expand in the future. Energy resource availability, climate change, air quality, energy security, and independence from other nations to maintain our standard of living will likely have a combined impact on the allocation and use of a variety of energy sources. The questions in this scenario are what role nuclear power will play and whether there will be the political will and technical work force present to allow nuclear power to occupy an increasing percentage of future energy needs.

Currently, about 16 percent of the world's electricity is generated from 435 nuclear reactors in 31 countries. Most of these are thermal reactors, which use neutrons at very low energies to induce the fission reaction. This is in contrast to fast reactors, which use neutrons that are orders of magnitude more energetic. Over 80 percent of commercial thermal reactors are light water reactors (LWRs) that use normal hydrogen in water for neutron moderation. During neutron moderation, neutrons with energies on the order of several million electron volts lose their energy as a result of numerous scattering reactions with a variety of target nuclei. After a number of collisions with nuclei, the neutrons have energies the thermal range (~0.025 eV).

There are many advantages to nuclear power. In a fission event, 200 million eV (10^{-11} J/atom) are released per atom that undergoes fission. In contrast, the breaking of chemical bonds in fossil fuels produces approximately 10^{-19} J/atom. Therefore, nuclear energy produces 10^8 times more energy per atom than fossil fuels. Additionally, nuclear fuel has a limited use to mankind, while fossil fuels have a wide variety of applications, including pharmaceuticals, plastics, and petroleum. Nuclear power does not produce greenhouse gases like fossil fuel burning plants.

For full public acceptance of nuclear power, a number of issues must be addressed, including waste disposal, reactor safety, economics, and nonproliferation. All of these issues depend on the fuel cycle that is used, but for any fuel cycle a geological repository will be needed for high-level waste storage. What will differ are the nature, hazard, half-life, and volume of the waste.

Of a 1000-kg block of uranium oxide fuel going into an LWR, only about 3.5 to 4.0 percent is fissile material (i.e., 3.5 to 4 percent ^{235}U or 35 to 40 kg) and the remaining mass is ^{238}U. After about 2 to 3 years, the fuel is spent and must be replaced. Of the original ^{235}U, about 75 percent has reacted by capturing a neutron and fissioning to produce a wide variety of fission products with half-lives that range from very short (days to tens of years as in the case of ^{90}Sr and ^{137}Cs) to very long (hundreds to millions of years as seen with ^{99}Tc and ^{129}I). Products heavier than uranium are also made by the capture of neutrons that do not lead to fission. In general, the unreacted uranium isotopes represent about 95 percent of the LWR spent fuel; the fission products, about 3.4 percent; all of the plutonium isotopes, about 1 percent; and the minor actinides (primarily neptunium, americium, and curium), about 0.6 percent. If the spent fuel is chemically processed to recover the unused uranium and plutonium, only about 4 percent of the spent fuel (the minor actinides and the fission products) needs to be managed as high-level waste.

There are three major fuel-cycle options that are being studied worldwide. The first is the once-through fuel cycle. In this option the fuel in an LWR is used for 2 to 3 years. It is then successively put into wet storage, dry storage, and eventually into a geological repository. No material is recovered or recycled; therefore, in the short term this power production method is the most economical since no costly chemical processing plant is required. It is also the safest regarding proliferation potential because the fissile material the fuel is "protected against diversion" by an intense radiation field. However, in the long term this option yields a very large amount of waste destined for the repository. In addition, after 250 to 500 years, a radiation field will no longer protect the fissile material because the fission products will have mostly decayed away. Because this option does not remove uranium and plutonium prior to geologic disposal, criticality issues must be addressed in the design of the repository, and the need for multiple repositories over time must be considered.

There are many challenges to the chemical sciences with the once-through fuel cycle. Scientists need to provide a scientific basis for understanding the performance of the geological repository. People must be convinced that the risk to the public is low and acceptable. Finally, chemists must determine the relevant physical and chemical processes occurring over the time the material is in storage. These characteristics and processes then need to be measured and used to validate models that assess the performance of repositories over thousands of years.

The second option is the reprocessing fuel cycle. Unlike the once-through option, with reprocessing the fuel cycle is actually closed since the unused fissile materials ^{239}Pu and ^{235}U are recovered and recycled back through a reactor to produce more energy. Before ^{235}U can be reused as a fuel and sustain criticality in an LWR, it has to be enriched back to the level of 3.5 to 4.0 percent. Plutonium, on the other hand, can be recycled by mixing it with either natural or depleted uranium to make mixed oxide fuel and burned in either a thermal or a fast reactor (a reactor that uses no moderator and induces fission with neutrons at energies at or near where they were created, ~2 MeV). Because the used ^{235}U is recovered, reprocessing not only allows a higher fraction of the energy content of uranium to be utilized but also reduces enrichment costs. Reprocessing also offers a convenient method of separating out the wastes (fission products and minor actinide elements) and placing them into a form acceptable for a waste repository. Currently, these waste materials are vitrified, that is, put into a boron-silicate glass.

The industry standard for reprocessing is the PUREX process (**P**lutonium **U**ranium **R**edox **EX**traction), which yields uranium and plutonium in forms suitable for making new fuel. PUREX is an aqueous-based solvent extraction process that, unfortunately, creates a large amount of high-level liquid waste that contains the majority of the fission products and the minor actinides. This waste has to undergo denitrification and calcinations before it can be vitrified. Additionally, other waste streams are created during the process because solvents and reagents undergo radiolysis and degrade as a result of the intense radiation. Chemists and chemical engineers are challenged to improve the PUREX process in modern plants, to improve the selectivity of reagents and optimize the reaction kinetics, to minimize solvent degradation and improve the recycle of solvents and reagents, and generally to reduce the number and volume of secondary waste streams.

In the reprocessing fuel cycle, the initial motivation for the fast reactor cycle was to "breed" plutonium from uranium to maximize uranium fuel utilization. This is because less than 1 percent of the energy content of mined uranium is realized in burning uranium in conventional thermal LWRs. In fast reactors, ^{239}Pu is made by placing a blanket of uranium around the core, where ^{238}U captures a neutron to produce the short-lived ^{239}U ($t_{1/2}$ = 23.5 minutes). ^{239}U then undergoes beta minus decay to another short-lived isotope, ^{239}Np ($t_{1/2}$ = 2.35 days), which again undergoes beta minus decay to finally become ^{239}Pu. Fast "breeder" reactors are configured to produce more ^{239}Pu than they consume. As a result, since the natural abundance of ^{238}U is 99.3 percent and the conversion efficiency through breeding is on the order of 70 percent, an almost infinite supply of fissile fuel for future generation could be produced using fast reactor technology. Because the expected shortage of uranium never materialized and uranium remained abundant and cheap in the mid-1980s, the need for fast reactors to breed plutonium was no longer compelling or economical. This, along with mounting concerns about the potential proliferation of plutonium, caused interest in fast reactor technology development to further wane. Recently, fast reactors are gaining interest as an efficient means of destroying long-lived actinide elements in more advanced fuel cycles.

The last and most futuristic fuel-cycle option is an extension of the reprocessing fuel cycle. In this option, in addition to recovering uranium and plutonium to produce more energy, the long-lived minor actinides and some of the long-lived fission products are chemically separated (partitioned) from the high-level waste stream and then converted into less radiotoxic or otherwise stable isotopes by a process called transmutation. Transmutation uses nuclear reaction, or a process to transform a radioisotope into one that is stable or has a shorter lifetime than the original isotope. Transmutation can be accomplished through a combination of LWRs, fast reactors that are configured to burn plutonium rather than breed it, and subcritical accelerator-based transmutation devices. The waste stream remaining after partitioning would contain only relatively short-lived fission products that would go to a repository and decay to the background level of high-grade uranium ore in about 250 years.

The partitioning of neptunium, iodine, and technetium is already possible through modifications to the PUREX process. This is not the case for americium and curium. Other aqueous separation schemes involving more selective extractants are needed to separate americium and curium, because they are similar in charge and size (ionic radius) to the lanthanide elements. Also, the lanthanide elements constitute about one-third of the total mass of all of the fission products in the PUREX high level waste stream. For such separations to be economically viable, new methods cannot generate more secondary waste than is generated using currently accepted processes. Further, new reagents used in the separation must be synthesized in sufficient quantity at reasonable cost to be used on an industrial scale. Reagents and solvents must also be compatible with the solvent extraction and other process equipment being used with PUREX.

Probably the most inviting and challenging R&D area impacting waste management for the advanced fuel cycle is the development of nonaqueous processes. Examples of these include separations based on differences in volatility, liquid-liquid extractions involving molten salts or liquid metals, and electrochemical methods (electrorefining) that use a cell potential to selectively remove a metal by reduction on to a cathode. Because of the higher fuel burnup needed for transmutation and the need for multirecycling with minimum cooling time between fuel discharge and reprocessing, nonaqueous processes offer several important advantages. Nonaqueous processes are much more radiation resistant compared with aqueous processes and thus can be used to reprocess spent fuel after considerably shorter cooling periods. Other advantages of nonaqueous processes include the potential to produce small volumes of secondary waste, waste that is more suitable for long-term disposal, ease of reagent recycle, compactness of equipment, and reduced costs due to a smaller required plant footprint. At the present time, major drawbacks are smaller separation factors requiring multiple stages to achieve the required level of decontamination, limited throughput because non-aqueous processes are usually operated as a batch rather than a continuous process, and the need for highly controlled atmospheres to avoid unwanted side reactions such as hydrolysis and precipitation reactions.

When compared with fast reactor technology, accelerated-driven systems (ADS) compared with fast reactor technology are in their infancy. An ADS consists of a high-power proton accelerator, a subcritical target that produces a high neutron flux upon bombardment with high-energy protons, a blanket system that utilizes this intense source-driven neutron flux to fission the actinides and transmute some of the long-lived fission products, and a supporting chemical processing plant. The heart of the system is the subcritical target, where an intense (>100 mA), high-energy (1 to 2 GeV) proton beam impinges on a high Z target to produce neutrons in a process called spallation. Accelerator systems producing 40 or more energetic neutrons per incident proton are being proposed. The waste material is loaded into the blanket where it is "incinerated" by capturing one of the neutrons, causing it either to fission or to be converted to a short-lived or stable isotope. This process therefore reduces the radiotoxic inventory of the waste material. The heat produced from the transmuter (the combination of the target and blanket) can then be used to generate power, ~10 percent of which can be used to run the accelerator; the remaining 90 percent can be put on the power grid. Unlike fast reactors that operate on the criticality principle, where the nuclear reaction is sustained, ADS are subcritical. When the proton beam is off in an ADS, no neutrons are created and no nuclear reaction occurs.

Some of the challenges to the chemical sciences (also metallurgy and materials science) related to the ADS involve the design and operation of the neutron source as well as the form and chemical composition of the blanket. Nuclear chemistry and chemical and nuclear engineering expertise is of great importance for developing and implementing the processes needed in the supporting chemical plant. In the chemical processing plant, separation processes are used to first partition the spent fuel to obtain the waste material that is introduced into the ADS and then later to support multiple recycles, fuel fabrication, and finally production of the final waste forms acceptable to a geological repository.

The training situation is dire in nuclear chemistry, radiochemistry, and nuclear engineering. There is great concern that our nation will neither have the right expertise nor enough expertise to meet future demands. Therefore, it is necessary to immediately reinvest in the education system, particularly for the training of nuclear scientists. Otherwise, the United States could lose on many fronts—its leadership in nuclear science and technology, its ability to influence Third World or emerging nations, and its ability to safely manage the existing nuclear enterprise. Without a properly trained work force, in the future the United States will not be able to preserve its options involving nuclear technology and nuclear energy.

Issues related to the expansion of nuclear power range from waste management to nonproliferation. There are some technical solutions, but these solutions

> **Sidebar 8.1**
> **Nuclear Waste Management**
>
> Nuclear power is the only alternative to fossil fuels that is currently competitive on a cost per kilowatt-hour basis. In addition, nuclear power has the potential to minimize many of the problems associated with the use of fossil fuels, including greenhouse gas emissions, emissions of other pollutants, and potential supply problems.
>
> A key consideration in the expanded use of nuclear power is nuclear waste management. Every nuclear power plant produces both high- and low-level radioactive materials. The basic fuel of a nuclear power reactor contains uranium-235. The splitting of heavy atoms such as uranium during reactor operation creates radioactive isotopes of a wide range of lighter elements as fission products. Fission products produced in high yield and with half-lives in the range of 30 to 50 years such as cesium-137 and strontium-90, account for most of the heat and penetrating radiation in high-level waste over the first several hundred years. These high-level wastes are capable of producing lethal doses of radiation to humans during short periods of exposure. In addition to these byproducts, low-level radioactive wastes (in general considerably less radioactive but more prevalent in total volume than high-level wastes) are created. Examples of low-level waste include paper, rags, tools, and clothing contaminated with small amounts of mostly short-lived isotopes. The ability to effectively manage both high- and low-level wastes is a key component to the use of nuclear power to supply a substantial portion of the nation's energy needs.

need to be improved and new ones need to be developed. The greatest challenge may simply be to educate the public and increase their confidence in nuclear technology. Regardless of public opinion, the United States has long-term nuclear issues that require expertise in nuclear science, and chemists and chemical engineers need to ensure that we have that expertise.

While most low-level waste can be managed at the site where it is produced, there is an increasing call for consolidation of high-level waste in a geological repository, where these materials are to be stored for thousands of years. High-level waste is currently stored at dozens of nuclear facilities throughout the country as both unprocessed spent fuel and high-level liquid waste resulting from fuel reprocessing. Storage of high-level liquid wastes at some of these sites has been plagued by problems of leakage of radioactive materials from aging storage tanks. The centralized geological repository at Yucca Mountain in Nevada can hold up to 63,000 tons of commercial nuclear waste, in a facility that has been deemed by the U.S. Department of Energy to be geologically stable for thousands of years. Current U.S. policy calls for storage of primarily unprocessed spent fuel from commercial power production as well as some vitrified wastes in large corrosion-resistant canisters at Yucca Mountain.

Balanced against these calls, however, are criticisms of this plan for centralized storage. Critics contend that thousands of shipments of highly radioactive material will be required over decades and that when Yucca Mountain reaches capacity, tens of thousands of tons of high-level waste must still be stored where it was generated. In addition, some have questioned the stability of Yucca Mountain as a geological repository, claiming the site sits on both an earthquake zone and volcanic zone.

Particularly in light of the recent presidential approval of the Yucca Mountain repository, it is clear that effective management of high-level waste will continue to be a key component in any plan to expand the nation's use of nuclear power as a key component of its energy needs.

9

Materials Technologies for Future Vehicles

Kathleen C. Taylor and Anil Sachdev,
General Motors Corporation

The motivation for new materials in the automotive sector falls into three categories—government regulations for emissions and fuel economy, technological advances, and market trends. Government regulations for emissions and fuel economy have had a large impact on the introduction of new lightweight materials and new catalyst systems. Technological advances can and do occur in reaction to government regulation; however, this is not always the case. For example, in California the requirement for electric vehicles did not lead to long-range, low-cost battery systems. Instead, hybrid power train systems are being developed that address energy efficiency with acceptable vehicle range. The market, particularly through consumer demand, provides tremendous challenges and opportunities for the automotive industry.

In response to these demands, improved energy efficiency is going to come from three areas--vehicle mass reduction, changes in basic vehicle architecture, and the power train. While the focus of this presentation is materials, improved efficiency of the power train plays a large role in energy savings.

The development and use of new materials are crucial to a response to the need for improved energy efficiency. In the year 2000, 17 M tons of iron and steel, 2 M tons of aluminum, and 2 M tons of plastics were used in the automobile industry. However, over the past 20 years, the allocation of these materials in a typical family car and the addition of new materials, have considerably altered the materials composition of automobiles (Table 9.1). Iron and steel, which made up 75 percent of a vehicle in 1977, have been reduced to 66 percent, while high-strength steel has risen from 3 percent to roughly 10 percent. In addition, the amounts of polymer composites and aluminum used have risen substantially.

TABLE 9.1 Materials Used in a Typical Family Car

	1977	1999
Total iron and steel	75 percent	66 percent
High strength steel	3.4 percent	10 percent
Polymer composites	4.6 percent	7.5 percent
Aluminum	2.6 percent	7.2 percent
Magnesium	0 percent	0.2 percent

Magnesium, which was virtually unused in automobiles in the 1970s, now makes up 0.2 percent of the typical family car.

Figure 9.1 shows fuel consumption versus curb weight for a large number of vehicles. The slope indicates that fuel consumption generally is greater as curb weight increases. However, achieving a reduction in vehicle weight by reducing vehicle size reduces functionality and contradicts market trends. To both provide the volume of space that consumers want and meet fuel economy standards requires the use of new materials and new power plants for vehicles.

For the automotive industry, tremendous challenges and opportunities are associated with introducing new materials, in terms of both fundamental and applied research. Issues related to performance/function, total accounted cost, design rules, and manufacturing feasibility at production volumes all play key roles. Other issues include joining (welding, bonding, and fastening), lead time for qualifying new materials, durability and reliability, and crash energy management.

The performance of a material in its intended application is the first materials requirement. For example, to convert from steel to aluminum for the vehicle body, differences in materials properties must be well understood and new methods must be developed for materials processing and manufacturing.

Design rules are different as a material is changed. When changing to a new material, new design guidelines for manufacturing are required in order to achieve high-rate and high-quality production.

As new polymer composites are considered, the issues of environmental impact and recycling become even more important. Metals have an advantage over polymer composites because an infrastructure exists for recycling them. However, work is progressing on developing more recyclable polymer composite systems.

Aluminum is the leading new material for lightweight vehicle structures. In order to be used more widely in passenger vehicles, a number of issues must be addressed. Forming and joining are of particular importance. Aluminum should deform better than steel based on its face-centered structure, but this is not the case. As a result, joining is difficult with aluminum. Galvanic corrosion presents

Sidebar 9.1
Materials Chemistry in Energy and Transportation

The making of materials requires carrying out numerous complex chemical reactions. Successful methodologies use various combinations of organic chemistry, solid-state chemistry, organometallic chemistry, inorganic chemistry, colloid and surface science and coordination chemistry, combined with chemical engineering. Namely, materials chemistry is derived from all the key areas of chemistry and chemical engineering that emphasize synthesis in order to develop new materials with the desired properties for the intended application.

The chemical sciences, especially synthetic chemistry, play an important role in four key areas with respect to energy and transportation. These areas are polymers and composites, catalysts and adsorbents, membranes, and photonic materials. Machinable ceramic parts and high-strength polymer composites have changed automobiles, allowing design and construction modifications that have led to the hybrid vehicles of today. Catalysts are used in scores of chemical and petroleum refining operations. Modern homogeneous and heterogeneous catalysts stabilize site-specific atomic clusters and molecular species capable of selective bond-breaking and bond-making steps. Selectivities to noncarbon dioxide species of about 90 percent are possible in butane oxidation, propylene oxidation, propylene ammoxidation, and ethylene oxidation. In the future, managing the carbon losses will become increasingly important. Adsorbents that include carbons, microporous aluminosilicates, and newer nano- and mesoporous hybrid organic-inorganic solids have completely changed the field of high-purity gases and separations—an industry that in the past relied on cryogenic processes but now has largely converted to adsorption-based processes with considerable savings in energy costs.

In response to other needs in the energy and transportation sector, membranes are evolving that transport molecular species, ions, electrons, and combinations of these species. For example, mixed oxide ion-electronic conductors that become the wall of tubular reactors will soon move out of the laboratory and be used to oxidize methane by transport of oxygen from the air side to the fuel side, where the methane is converted to carbon monoxide and hydrogen. This technology eliminates the need for huge and expensive air separation plants to supply oxygen.

Solar-based strategies for the conversion of the sun's light to electrical and chemical energy required the discovery of a new generation of specifically designed and engineered photonic materials. Examples of these inorganic semiconductors include those based on amorphous silicon, cadmium telluride, and more complex ternary or quaternary combinations such as copper iridium diselenide or dye-sensitized semiconductors.

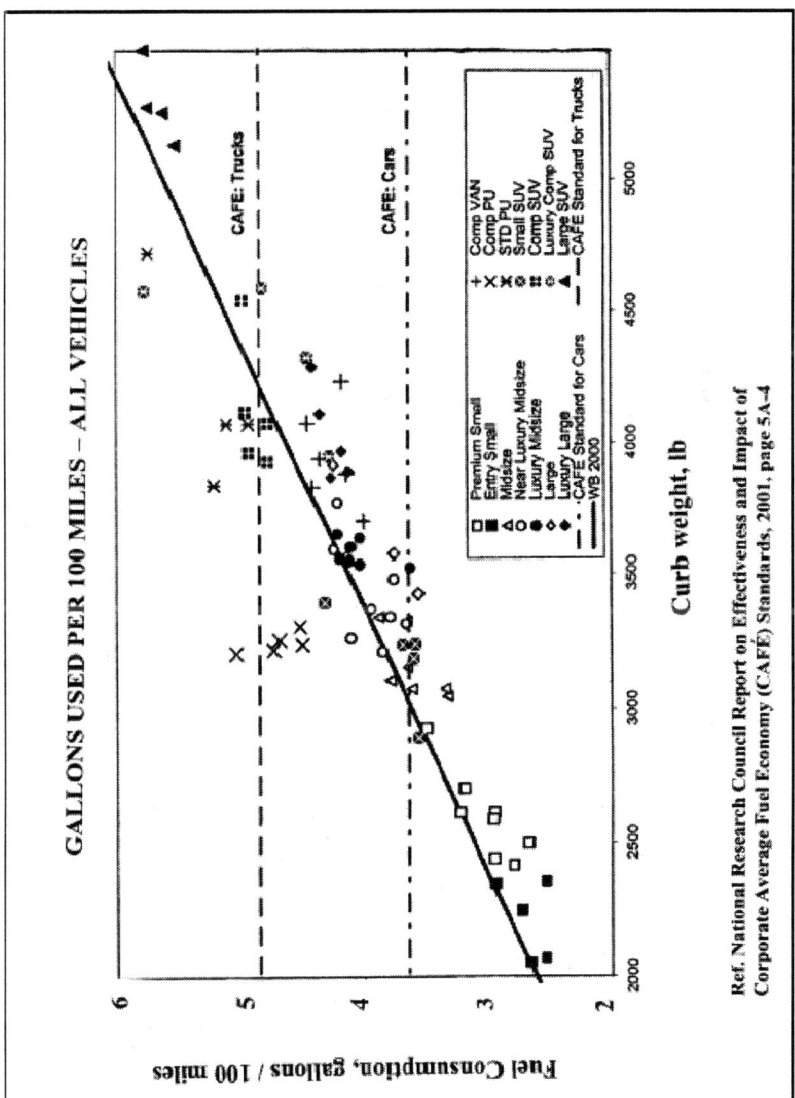

FIGURE 9.1 Fuel consumption versus curb weight.

another challenge because it restricts the manufacturer to medium-strength alloys. Very high strength alloys pose problems of stress corrosion and cracking.

Limits on aluminum alloy composition are mainly driven by recycling concerns. Aluminum cannot be readily refined to remove secondary elements without added cost. As a result, when developing composites, auto manufacturers try to limit applications to one or two alloy systems.[1]

Painting can be an issue when using aluminum because of formation of the aluminum oxide layer. Because the modus of aluminum is one-third that of steel, aluminum parts must be a little thicker than steel in order to resist denting.

General Motor's development of the Oldsmobile Aurora represents a vehicle that uses aluminum extensively. Aluminum is used in the cylinder head and cylinder block, the hood deck lid, the fenders, and the bumper beams. The use of aluminum in these applications has reduced the weight of this particular vehicle by 285 pounds. An aluminum engine cradle was used for the first time in the 2000 Impala. The engine cradle is made from 19 extrusions and one stamping, which are all put together in a large welded fixture for a weight savings of 20 pounds. GM plans to produce 1.5 million units of this engine cradle.

Magnesium is another material that has increased in use. One large application for magnesium in automobiles has been the instrument panel for GM's full-sized car and full-sized van that consists of a one-piece die casting that provides the entire structure of the front dashboard.

The use of polymer composites in automobiles poses many advantages as well as significant barriers. Polymer composites offer low mass and are damage and corrosion resistant, and parts can be consolidated because of molding capability. Balanced against these benefits is the significant barrier in changing current assembly methods. For the manufacture of auto parts from composites, entirely new assembly methods must be developed. Issues that must be addressed include the durability of composites, design issues, and the cycle time for manufacturing.

GM recently put into production a new nanocomposite made from thermoplastic olefin with a clay filler. Only about 2 percent clay is used, compared with 20 to 30 percent talc in the composite material it replaces. Two advantages of this new nanocomposite are that it is both light-weight and recyclable. Presently, this nanocomposite is used in the step assist for the full-sized van, but a future goal is to use it in body applications such as vertical and horizontal panels.

Polymer composites are also being used for the pickup box in trucks. Using a composite in this application saves 50 pounds with the truck bed, and another 15 pounds with the lift gate and provides superior impact and corrosion resistance.

In the past decade, interindustry-government partnerships such as the Partnership for a New Generation of Vehicles (PNGV) have become a significant means

---

[1] In addition to recycling costs, the costs of aluminum stampings have also placed limitations on additional use of aluminum alloy composites in lightweight vehicles.

for conducting precompetitive research. Reduced vehicle mass was recognized as a key factor for improving fuel economy in the PNGV program, and a number of companies have each used an aluminum body structure and a light power train to achieve this goal in their year 2000 PNGV concept vehicles.

The steel industry is also looking into new ways to utilize high-strength steel in the construction of motor vehicles in order to reduce vehicle mass. Other developments are steel-plastic sandwich structures and innovative manufacturing processes. The American Iron and Steel Institute plans an advanced concept vehicle project targeted to meet the 2004 PNGV vehicle and crash requirement.

Future needs are divided into three areas—lightweight materials, low-cost materials and processes, and environmental enablers. Light-weight materials provide fuel economy while keeping the vehicle volume constant and maintaining the power of the vehicle for activities like towing. The introduction of new lightweight materials must be done without increasing the base cost of the vehicle, and high-volume manufacturing methods must be used that produce high-quality materials.

10

Could Carbon Sequestration Solve the Problem of Global Warming?

Stephen W. Pacala,
Princeton University

During the 21st century, it is anticipated that a trillion tons of carbon in the form of carbon dioxide from anthropogenic sources will be emitted into the atmosphere.[1] While it is uncertain what the long-term effects of this added carbon will be, ecological models indicate that a significant amount of damage to ecosystems could result.

In addition to increased use of renewable carbon-neutral energy, an additional "backstop" technology to decrease the amount of carbon emitted into the atmosphere is to sequester it. The key question for carbon sequestration technology is: can the science and technology be developed to effectively remove carbon from the atmosphere and keep it sequestered? To answer this question, global biogeochemical constraints must be addressed.

Research has been done on so-called natural biological sinks to determine the amount of carbon sequestration possible by this method. This work has indicated that natural biological sinks such as a pine forest do initially achieve elevated carbon dioxide fixing from the atmosphere but that this effect goes away after approximately 3 to 4 years. This effect, termed down regulation, can be the result of a long-term decrease of nitrogen or other needed nutrients in the forest over time.

A global extension to this question is whether the world is as a whole down regulated or if carbon dioxide fertilization is actually occurring. U.S. Forest Service data for the past 70 years was analyzed to find if forest growth is currently

---

[1]National Research Council, 2001, *Carbon Management: Implications for R&D in the Chemical Sciences and Technology*, National Academy Press, Washington, D.C., pp. 8-14.

faster than in the past, indicating that carbon dioxide fertilization is occurring. The results indicated that growth was exactly the same presently as it was when the records were first kept. Extrapolating these data globally, the conclusion is that the world has indeed down regulated. However, a careful inventory of U.S. data indicates that, although the country is taking up a half billion tons of carbon dioxide annually, this is almost entirely the result of recovery from past land use. The problem with the land-use sink for carbon is that eventually the sink goes away.

In addition, the extent of increasing anthropogenic carbon in the atmosphere over the 21st century is so extensive that the contribution that land use carbon sequestration could make is probably quite limited anyway. As stated previously, it is anticipated that a trillion tons of carbon in the form of carbon dioxide from anthropogenic sources will be added to the atmosphere during the 21st century. Assuming that all of this added atmospheric carbon must be removed, conversion of all agricultural lands and grasslands globally into old-growth forests would remove only 475 billion tons.

A second possibility for a carbon sink is the world's oceans. At present, there is already some 37 trillion metric tons of carbon, mostly in the form of bicarbonate, dissolved in the oceans of the world. Of the carbon not taken up by terrestrial ecosystems, the oceans will be the eventual repository for about 85 percent of the rest of the carbon emitted to the atmosphere by human activities.[2] However, this uptake occurs quite slowly. For example, the oceans are currently taking up only 40 percent (with an uncertainty of ±16 percent) of the annual anthropogenic carbon emissions not removed by terrestrial processes. Because of the slow rate of mixing of the world's oceans, it would take many centuries for them to realize most of their uptake capacity, even if anthropogenic emissions were to stop today.

The oceans capacity is such that all anthropogenic carbon dioxide can be absorbed. However, the problem faced with the carbon cycle is that this anthropogenic carbon dioxide is put into the atmosphere faster than the oceans can absorb it.

Extensive modeling on the absorption of carbon dioxide by the oceans has been confirmed through carbon-14 penetration in the oceans from atmospheric testing of nuclear weapons. Using these models, it is possible to predict methods to artificially remove carbon dioxide from the atmosphere, such as injecting it directly into the oceans as a gas or as supercritical carbon dioxide. The results indicate that significant problems result from this approach. First, injecting carbon dioxide into ocean shallows results in escape back into the atmosphere. To overcome this obstacle, carbon dioxide must be injected deep into the ocean abyss,

---

[2] P.,Tans, J.L. Sarmiento, and W.H. Schlesinger, 1998, "Changes in Carbon Sources and Sinks: The Outlook for Climate Change and Managing Carbon in the Future," USGCRP Seminar 8, December, Washington, D.C., http://www.usgcrp.gov/usgcrp/seminars/981201DD.html.

which drives up costs astronomically. In addition, localized injections of massive amounts of carbon dioxide into oceans are likely to cause significant ecological damage, through localized changes in pH and inhibition in the growth and reproduction of deep-sea animals. Therefore, a number of smaller injections at different points in the oceans must be done, which also dramatically drives up costs. Finally, the long-term ecological effects of such a scenario, such as formation of clathrates that would smother sea bottom organisms, are either unknown or detrimental.

Other methods proposed for ocean sequestration may lower costs but present significant obstacles. One scenario envisions fertilization of the oceans with iron. However, studies of this method have indicated that only about 10 percent of the carbon fixed remains in the ocean. Also, the unintended ecological consequences of this method, such as abyssal nitrogen anoxia and risks to fisheries, may far exceed the benefit derived from carbon sequestration.

Geological sequestration poses another possibility to fix atmospheric carbon. Presently, the oil and gas industries collectively move hundreds of millions of tons of gases, including carbon dioxide, and re-inject them into fossil-fuel-bearing geological formations either as waste gas or to enhance oil recovery as part of their normal operations.

The key problem with this method is that these geological reservoirs potentially leak, either through natural fractures or by puncturing from hundreds of thousand of old wells that are sealed with concrete caps that have the potential to leak. Backstopping technologies, such as the formation and burial of carbonate rocks from carbon dioxide, have been proposed to overcome this limitation. The impediment to this method is the cost effectiveness of moving and burying these large rocks.

Although leakage of carbon dioxide from these geological repositories is a concern, it is not necessary to completely seal all of these leaks to effectively avoid potential climate changes due to added carbon dioxide in the atmosphere. Mathematical modeling can be performed to balance anthropogenic carbon dioxide generation, geological carbon sequestration, and leakage of carbon dioxide from these reservoirs back into the atmosphere. Results indicate that geological sequestration has the capacity to solve the problem of excess anthropogenic carbon dioxide in the atmosphere, provided that the leakage rate from these repositories is kept beneath 1 percent per year.

There are key challenges for the chemical sciences if sequestration is adopted as a method to reduce atmospheric carbon dioxide. Carbon dioxide capture after generation is generally estimated to represent three-fourths of the total cost of a carbon capture, storage, transport, and sequestration system.[3] To make carbon

---

[3] http://www.fe.doe.gov/coal_power/sequestration/sequestration_capture.shtml

sequestration practical, research in the chemical sciences in the following areas will be required:

- absorption (chemical and physical),
- adsorption (chemical and physical),
- low-temperature distillation, and
- gas separation membranes.

In addition, more information is needed about the chemistry of carbon dioxide in brine with mineral surfaces. Chemical tracking of carbon dioxide is required to determine how to plug leaks without eliminating the storage capacity of a reservoir. Also, understanding how to keep gases (such as hydrogen sulfide from coal) in solution that would cosequester with carbon dioxide (which would also leak from reservoirs but unlike carbon dioxide would cause a substantial localized problem) is also a fundamental problem to be solved by the chemical sciences.[4]

---

[4]It is important to note that any sequestering process requires energy. For sequestering to be a useful process, this energy cannot produce carbon dioxide, or at least should produce much less than is being sequestered.

11

The Hydrogen Fuel Infrastructure for Fuel Cell Vehicles

Venki Raman,
Air Products and Chemicals

For hydrogen fuel to be a viable fuel for transportation, an extensive hydrogen fuel infrastructure must first be built. In assessing whether such a network is desirable, the benefits and challenges associated with introducing hydrogen as a fuel must first be addressed. In addition to infrastructure issues, hydrogen production on an industrial scale must be examined, and the changes needed for it to become a mass market fuel must be addressed.

There are many positive factors related to hydrogen fuel. Most technologists involved in the development of fuel cell vehicles realize that hydrogen fuel offers lower vehicle costs due to the simpler designs and fewer pieces of equipment onboard the automobiles as opposed to reforming gasoline to make hydrogen on the vehicles. While vehicle fuel storage designs are not optimal, existing systems for compressed (high-pressure) hydrogen are operable. Once hydrogen is on a vehicle, there are no pollutants emitted, only water vapor. Finally, using fossil fuels to produce hydrogen allows maximum energy flexibility—the ability to switch from fossil fuels today to renewable starting energy sources in the future, since the infrastructure would exist.

Unfortunately, the lack of a hydrogen infrastructure presents one of the main challenges to introducing hydrogen into the mass market as a transportation fuel. The nation has already seen such a scenario with natural gas—it is widely available, has adequately developed technology, and is economical but natural gas has not penetrated the market. There is also the perception that hydrogen is explosive and unsafe, which needs to be countered by public education. High-pressure gaseous hydrogen or cryogenic liquid hydrogen are inconvenient for individuals to use, and hence better hydrogen storage solutions are required. One desirable option involves solid-state systems operating at low pressure and ambient tem-

> **Sidebar 11.1**
> **Hydrogen Production and Storage**
>
> Low-cost production of hydrogen from water and its utilization on a massive scale as an energy carrier, substituting for fossil fuels, would constitute a major advance toward alleviation of the build up of atmospheric carbon dioxide and its associated potential effect on global climate change. Hydrogen can be substituted for carbon-based fuel for transportation, for electric power generation using fuel cells, and for process heat. For transportation applications, hydrogen can be used to drive a fuel cell power train or used directly in an internal combustion engine; the former eliminates all undesirable emissions. The challenge for chemistry is to produce and store hydrogen gas at normal atmospheric conditions, at sufficiently low cost and high density, to make its utilization as a transportation fuel as well as its use for other energy applications economically feasible.

peratures. Inadequate hydrogen storage options may be the greatest barrier to mass hydrogen use.

Fleet applications can certainly lower the hurdles to the entry of hydrogen into the fuel market. Usually, bus fueling is done once a day at a central depot. The market can be readily served in the same fashion as existing industrial hydrogen market applications. In fact, there have already been several projects with fleets that have given us the knowledge to clear some hurdles to pave the way for the introduction of hydrogen into the fuel market.

Presently, 40 million tons of hydrogen are produced worldwide each year. Ninety-five percent of the hydrogen produced is used captively. That is, the hydrogen molecule is made and immediately reacted to refine oil or to produce ammonia, methanol, and other chemicals. Another 5 percent of the 40 million tons of hydrogen is produced by the merchant hydrogen market—including large industrial gas companies such as Air Liquide, Air Products, BOC, and Praxair—for sale to third parties. Eighty percent of the hydrogen produced is made from natural gas steam methane reformation. In addition, there are byproduct streams from which hydrogen can be captured and purified.

Hydrogen can be transported as a high-pressure gas or a liquid. The high-pressure gas can be moved economically only approximately 100 miles from the plant because of the weight and size of the metal cylinders needed to contain the gas. Although there are very few applications that require liquid hydrogen, it is the most convenient form for transporting hydrogen over long distances. Liquid hydrogen can be moved economically approximately 1,000 miles and is usually revaporized before use.

Hydrogen at high pressures can be conveyed from the production plant through a pipeline system to a consumer that can be right across the fence or miles away. Air Products operates several hundred miles of hydrogen pipeline in various parts of the United States and the world through which almost 2,000 tons of hydrogen flow every day. The pipelines are fed through multiple production plants, and numerous customers draw hydrogen from them.[1]

Market projections for fuel cell vehicles in 2015 indicate less than 5 percent penetration of fuel cell vehicles into the worldwide vehicle population. Theoretically, there would be approximately 150,000 buses and between 20 million and 80 million light-duty vehicles, which would consume between 20 million and 90 million tons of hydrogen per year. Worldwide hydrogen merchant capacity is presently 2.5 million tons per year. To meet these future fuel cell vehicle needs, either hydrogen must be produced in large central plants and delivered via one of the three previously mentioned modes (high-pressure gas delivery, liquid delivery, pipeline delivery) or very small on-site hydrogen production plants that use electrolysis or reforming technology must be built.

There are many more challenges to overcome. To give a car enough range to operate effectively, the pressure of hydrogen fuel must be between 350 and 700 bar—a pressure well in excess of those currently used in industrial practice. The ease and speed of fueling are not at levels sufficient for consumer demand. Because of the intermittent nature of refueling, hydrogen production and storage systems will be required, which are significant cost items in any infrastructure design. Hydrogen flow will need to be metered for payment, but current technologies need to be improved.

There have been a number of demonstration programs that have been conducted for fuel cell vehicles, all based on delivering liquid hydrogen to the fuel station compression and dispensing it to vehicles. A successful program with the Chicago Transit Authority used a reciprocating liquid pump, which can supply a vehicle at very high pressures. Other demonstration programs at SunLine Transit Agency in Palm Springs and Coast Mountain Transit in Vancouver, Canada, used water electrolysis to produce hydrogen.

When fuel cell vehicles begin to be introduced, there will be a very small number that are geographically widespread. This will in turn require small-capacity hydrogen plants, and the sporadic demand will result in inefficient capital utilization and high costs. A new project starting in September 2002 in Las Vegas will test the viability of baseloading the hydrogen plant with a fuel cell power plant and using incremental hydrogen capacity for fueling operations. This

---

[1]If this method is to be considered for hydrogen fuel vehicles, the economics and efficiency of liquefying hydrogen must be taken into consideration.

"Energy Station" concept will be run for 2 years to understand the economics of this approach.

In the early days of the hydrogen transportation market, demonstrations and start-up projects will be fueled by liquid hydrogen or distributed hydrogen generators, like those in development for Las Vegas. This could handle less than a ton of hydrogen per day. As demand grows, there will be larger on-site production plants and perhaps even regional plants that could handle 10 to 100 tons per day.

Today, world-scale hydrogen plants are in the size range of 100 to 250 tons a day. As demand for hydrogen as a fuel increases, chemists and chemical engineers will be presented with the opportunity to develop low-cost hydrogen production, especially for small-scale plants.

Hydrogen demand will grow by orders of magnitude when fuel applications expand. There are many challenges to using hydrogen as a fuel. Some are already being met by adapting existing capabilities for the demonstration programs that have been implemented. There is a logical and stepwise pathway to grow the hydrogen infrastructure by adapting industrial hydrogen production experience to hydrogen fueling demonstrations.

12

Opportunities for Catalysis Research in Energy and Transportation

R. Thomas Baker,
Los Alamos National Laboratory

INTRODUCTION

Catalytic processes, key to a variety of efficient transformations, are of enormous importance and are responsible in many ways for the high standard of living we enjoy today. Approximately one-third of the Gross Domestic Product of the United States today can be traced to the success of one or more catalytic processes. In the chemical industry approximately 80 percent of the processes use catalysts, and the percentage is expected to grow as concern over waste disposal grows. By definition, catalysis saves energy and plays a key role in diverse areas in energy and transportation. Future opportunities for catalysis in energy and transportation will rely on the development of more efficient catalytic processes that take place under milder conditions (lower temperature and pressure) along with the discovery and development of selective new catalysts for transformations that either are not known or are too inefficient today.

A catalyst simply provides a reaction pathway that is not possible in its absence. A perfect catalyst would catalyze a reaction forever with no activation energy and 100 percent selectivity. Although many catalysts have turnover numbers in the millions, most catalyst lifetimes are limited by side reactions that lead to catalyst deactivation. Modern study of chemical catalysis is an interdisciplinary enterprise that involves scientists from every chemical discipline collaborating with each other and with material scientists, theoreticians, and chemical engineers.

The largest area of catalysis in industry today is heterogeneous catalysis. Heterogeneous catalysts have the advantage of being insoluble and often relatively robust. Therefore, gaseous or liquid reactants and products readily pass through or over a heterogeneous catalyst, and the product can be separated from the catalyst readily. Disadvantages of heterogeneous catalysts include the inability to control activity by systematically altering the surface structure, the difficulty of identifying precisely the type of site that promotes a given reaction, and competing side reactions at other catalytic sites on the surface. It is also difficult to guarantee that a high percentage of a metal in a heterogeneous catalyst is participating in the catalytic reaction, thereby requiring either higher temperatures or high metal loading for practical operation.

Homogeneous catalysts, reactants, and products are all in the same phase, usually liquid. Homogeneous catalysts often employ metal complexes with well-defined ligand coordination environments and reactivities, which lead to a high percentage of active catalyst sites, and predictable and reproducible activity that depends on a 'single site' being present. The main disadvantage of homogeneous catalysts is separation of the product from the catalyst, a problem that has been solved in some cases using phase transfer catalysis, or by 'tethering' a homogeneous catalyst to an insoluble support. A second disadvantage is the widespread use of hydrocarbon solvents, which has led to an exploration of the possible use of other solvents (water, supercritical carbon dioxide, and ionic liquids) in many reactions.

In the future new metal catalysts and catalytic reactions will be perfected primarily as a consequence of fundamental studies of reactions at a metal center. One must be able to control the timing of a sequence of steps in the primary coordination sphere of a metal with exquisite delicacy and thereby direct the metal through a maze of possible outcomes, including no reaction at all. The challenge is to control the outcome of a desired catalytic reaction, either through ligand design in homogeneous catalysts or surface design in heterogeneous catalysts.

OPPORTUNITIES FOR CATALYSIS RESEARCH

Below are some examples of catalysts and areas of catalyst research that could contribute significantly in the future in terms of energy savings and improved transportation.

De-Nitrogen Oxide Catalysts

The transportation sector accounts for 68 percent of all of the petroleum used and one-third of the anthropogenic carbon dioxide emissions in the United States. In addition, the utilization of internal combustion engines for transportation results in a significant amount of nitrogen oxides and particulate emissions.

Technology exists today to manufacture much more thermally and fuel efficient 'lean burn' engines (compression ignition, direct injection engines such as diesels or homogeneous charge compression ignition engines) that operate under lean conditions. Existing noble metal-based catalytic converter technology for stoichiometric gasoline engines is not suitable for the oxygen-rich exhaust of a lean-burn engine, so there is no suitable existing technology for emissions control for the lean-burn engine. The scientific challenge is to catalytically reduce nitrogen oxides under oxidizing conditions over a wide operating temperature range, from below 200 °C up to 500 °C.

To achieve this goal, new catalysts are being sought that could affect the selective catalytic reduction of nitrogen oxide in the presence of oxygen by exhaust hydrocarbons or by introduction of ammonia from the hydrothermal decomposition of urea. Molecular sieve zeolites and other oxide supports, ion-exchanged with transition metals and other ions, are the best of the catalysts known to affect this chemistry. However, the mechanism of the selective catalytic reduction is still not understood. In addition to discovering catalysts of high activity, the challenge is to maintain the activity of the catalyst under conditions of high temperature, steam, sulfur, and the long operating times necessary to make these materials viable catalysts for use in lean-burn engines.

The past decade saw additional major advances in automotive catalysts. The amount of costly precious metals such as palladium, platinum, and rhodium in catalytic converters has been reduced by more than 20 percent through the use of precision coating onto increasingly sophisticated mixed metal oxide supports. To reduce existing "smog" levels, Engelhard's PremAir technology uses a platinum-based coating on the radiator and air-conditioner condenser to convert ground-level ozone into oxygen and carbon monoxide into carbon dioxide.

Clean Coal Technologies

The fact that a massive supply of coal still exists in this country argues for technologies that allow for more efficient use of coal and that burn coal with fewer emissions and less carbon dioxide. One of the technologies presently being pursued is to combine anaerobic hydrolysis of coal with the capture of carbon dioxide by calcium oxide. The thermodynamics of these combined processes indicate that coupling these reactions leads to a thermodynamically neutral overall reaction.

The benefit of this type of process is that all potential emissions could be handled at once. However, burning coal releases not just small molecule emissions but sulfur and heavy metals in addition to carbon formation and radioactive emissions. Therefore, it is necessary to perform in situ diagnostics to receive feedback on transient concentrations in order to maintain the most efficient coal burning.

New Catalysts for Refining Applications

Over the past 60 years catalysis has transformed the refinery from a distillation plant to a sophisticated chemical-processing plant. One major challenge facing the refining industry is the cost-effective manufacture of ultra low sulfur transportation fuels to enable vehicles to meet stringent reduction in emissions. To enable the production of less than 10 ppm of sulfur gasoline, a number of new catalysts and processes have been developed to selectively desulfurize part or all of the FCC (Fluid Catalytic Cracking) naphtha and preserve the olefins that provide octane value. There are a variety of selective desulfurization options available that preserve octane value. These include IFP's Prime G+, BP's OATS process (Olefin Alkylation of Thyophenic Sulfur), Phillips' S-Zorb process, and ExxonMobil's SCANfining process.

A breakthrough distillate hydrotreating catalyst, jointly developed by ExxonMobil and AkzoNobel, was recently commercialized and deployed in three refineries. This catalyst has three to ten times the activity of the most active current HDS/HDN catalysts, depending on pressure. This catalyst also exhibits novel and unique pressure sensitivity from 400 psi Hydrogen-2 pressure up to 2,000 psi Hydrogen-2. It represents a major advance in catalyst performance, composition, structure, and morphology. The product is stripped not only of sulfur but also nitrogen and achieves substantial aromatic saturation.

Short contact time FCC has significantly improved the yield and quality of naphtha. A major improvement in cracking catalyst design came about from the recognition of bifunctionality of FCC catalysts. An example of a new catalyst is one where detrital alumina deposited on the active zeolite doubled activity and increased selectivity. The ADA catalyst family has been developed in a joint research program by Grace Davison and ExxonMobil. Much higher product value has been generated by significant improvements in the ZSM-5 catalyst system to produce olefins in FCC units, specifically ethylene and propylene. Major improvements in lube hydroprocessing have come about through the use of the new MSDW-2 isodewaxing catalyst.

Selective Reactions of Methane and Other Alkanes

Currently a great deal of energy is expended to obtain starting materials, usually alkenes, for the chemical industry from petroleum or coal. These alkenes are then converted into polymers, alcohols, and other downstream chemical products. In view of the large reserves of methane (natural gas), it would be highly desirable to use it as a source of products for the chemical industry. Although methane can be reformed with steam and oxygen to produce a mixture of carbon oxide and hydrogen ('synthesis gas'), which is then reacted over a catalyst to produce a variety of hydrocarbons, this 'Fischer-Tropsch' chemistry involves high pressure and temperatures.

It would be highly desirable to be able to convert methane directly into value-added products, most importantly into methanol by direct oxidation with oxygen at low temperatures and pressures. Selective direct oxidation of methane to methanol would allow plants to be built at the sources of methane, thereby eliminating problems associated with transportation of methane. Methanol at present is the starting point for the synthesis of multicarbon commodity chemicals and can be converted into alkanes using zeolite catalysts. Significant progress has been made in selectively oxidizing methane to methanol with oxygen and platinum catalysts, although an efficient catalytic process has yet to be perfected. It also would be desirable to directly functionalize higher alkanes, for example to, alcohols, or to dehydrogenate them selectively to alkenes.[1]

Catalysts for Photocatalytic Water Splitting

One of the grand challenges is to develop catalysts that will split water into hydrogen and oxygen using sunlight as the source of energy for this uphill reaction. This would be the artificial counterpart of photosynthesis in green plants. Hydrogen and oxygen could then be recombined in fuel cells, thereby effectively utilizing the energy of the sun to provide electricity with an efficiency that is not inherently limited theoretically to a small yield, as in silicon-based solar cells. Photocatalytic splitting of water is an ambitious project with a huge and important payoff. Complex new transition metal complexes must be designed that operate in water and allow the production and separation of hydrogen and oxygen using photochemistry. These catalysts are likely to be either homogeneous or tethered to a support that doubles as a hydrogen-permeable membrane, thereby allowing hydrogen to be separated from oxygen. The most ambitious approach would be to link photocatalytic splitting of water with a fuel cell in order to effectively convert sunlight into electricity in a single device. Although progress has been made in driving unfavorable reactions photochemically, we are not close to a solution to the problem. The solution will require long term support, research effort, and patience.

Linking Catalysis Theory with Experiment

With advancements in computer power and the advantage of reasonable scalability to larger systems, density functional theory has made computational chemistry widely accessible in the chemical sciences, permitting direct comparisons to be made between theory and a wide variety of experiments. Examples of the application of density functional theory for prediction, understanding, and interpretation in surface science and heterogenous catalysis include:

• understanding active site structures and structures of key intermediates in acid catalysis of hydrocarbon conversion by zeolites,

---

[1] As hydrocarbons will most likely be used for the foreseeable future, the development of hydrocarbon anode catalysis is another area of research that could potentially provide significant breakthroughs.

- understanding the importance of dynamic changes of the catalyst surface in ammonia synthesis,
- identification of new catalytic intermediates in selective oxidation;
- prediction of reaction energetics and the kinetics of selective and non-selective catalytic pathways,
- predictions leading to the discovery of a new Diels-Alder reaction for organic functionalization of semiconductor surfaces, and
- prediction leading to invention of new steam reforming catalyst.

Although computational chemistry is far from predicting the outcome of a chemical reaction through full-scale modeling of the catalytic reaction in question, it can be used in conjunction with experiments to guide the experimentalist in potentially fruitful directions. Efforts should be made to formalize collaborations between experimentalists and theoreticians as much as possible in the future.

13

Role of 21st Century Chemistry in Transportation and Energy

Jiri Janata,
Georgia Institute of Technology

CHEMICAL SENSORS

Chemical sensors, important for environmental and processing applications, are finding increasing utility in national security. National security requires constant vigilance for detection of the myriad chemical and biological agents that could potentially be used against the United States Because not every sensor can monitor the presence of all agents, the most recent and growing trend is to use higher-order chemical sensors and sensor arrays.

A chemical sensing array is achieved by obtaining multiple parameter measurements from one information channel in addition to using multiple channels. The information obtained then comes from "information space." This approach is similar to hyphenated techniques, such as gas chromatography-mass spectrometry, which separates components and then finds further information about those components with the second technique.

Chemical sensors are key in the transportation industry. Sensors are used to determine the fuel ratio, manage the optimum ratio, and measure the oil quality for pollution control. Chemical sensors monitor tailpipe emissions and catalytic converters. Additionally, sensors aid in chemical diagnostics of oil, transmission fluids, and other performance fluids.

Higher-order chemical sensing can alleviate some inherent problems of chemical sensors. For example, a sensor array can mathematically correct for systematic drift. It also provides cross selectivity for elimination of interference.

Sensor arrays are produced by microfabrication. Because different portions of the array have different fabrication requirements, convenience dictates that the sensors be fabricated in two parts: the chemical sensing chip with the transducer

Sidebar 13.1
Fundamental Discoveries that have Influenced the Development of Analytical Sensors

There are several classes of chemical and biochemical sensors that are increasingly indispensable in automotive, manufacturing, environmental, biomedical, and R&D applications. Most of these can be traced to seminal scientific advances that date back decades and, in some cases, to discoveries that occurred in the 19th century. Important connections between discoveries and classes of sensors are as follows:

- The *Nernst glower* is one of the oldest sensors and is used to measure the composition of automotive exhaust.
- Potentiometric sensors for analysis of solutions were enabled by the development of the *glass electrode*, and its coupling to a volt meter for measuring pH. This work was later extended to ion-selective electrodes using membrane materials such as Lanthanum Fluoride and *valinomycin/Polyvinyl Chloride*.
- Probably the most important developments in amperometric sensors were the *Clark oxygen electrode* and the amperometric *glucose sensor*. The latter is the most successful example of an enzyme-based sensor.
- Important discoveries in sensor platforms include the *tin oxide* "electronic nose," quartz crystal microbalance and related Surface Acoustic Wave devices, *field effect transistors*, and *optical fibers*.
- Molecular recognition has influenced sensor development, first with alkali ions and later with more complex host-guest interactions. New materials likely to have an important impact in the future are conducting polymers, semiconductor nanoparticles, and fluorescent and colorimetric reporter molecules.
- *Array chemical sensors* represent an important recent development that involves many sensor platforms. When combined with fast computational methods for pattern recognition and data analysis, these methods have substantial potential for analysis of complex mixtures and process control.

Miniaturization and integration of sensors into "lab on a chip" total analysis systems are likely to have very important implications in medical diagnostics and microanalysis of complex mixtures. The breakthroughs here have involved microfluidics and *laser-induced fluorescence* techniques, as well as *ultramicroelectrodes* for ultrasensitive electrochemical analysis.

> **Sidebar 13.2**
> **Sensors for Transportation Systems**
>
> One characteristic aspect of transportation is the high density of people in closed spaces. This may be in train stations, airport buildings, airplanes, subway systems, ships, and ground transportation media. Because access to all these spaces is wide open, they are vulnerable to a possible terrorist attack. To improve the security of transportation in general, new paradigms in chemical sensing and monitoring must be adopted. These include the rational integration of chemical sensors into protection models that take into account specifics of the close space to be protected. Introduction of distributed and dynamic modes of chemical sensing, wireless information communication links, and access to information databases must be exploited. In addition, artificial and rapid intelligence algorithms need to be developed. Hardware development will require new dedicated facilities for design and fabrication of integrated information electronics as well as the custom design of chemical-sensing arrays and higher-order chemical-sensing arrays.

that converts the primary interaction into a measurable signal and the data processing chip.

Conventional field effect transistors are the building blocks of binary electronics. Worldwide production is 10^{17} transistors per year, and per capita use in the United States is 10^7 transistors per year. The cost per transistor is one micro dollar, and the yield of fabrication is greater than 99 percent. The metalization on the interconnects of the chips was previously aluminum, but it is starting to be switched over to copper. Likewise, the solid-state circuit feature size was 0.1 μm but is now becoming smaller. Silicone dioxide comprised the top layer of the device.

On the other hand, chemical electronics (chemical sensing chip) production was probably less than 10,000 yearly (10^{-4} per capita) with a cost of $5 per chip. The yield is no better than 60 percent, and the metalization is platinum or gold. The feature size in the XY direction is typically on the micron scale, but it is 100 microns in the Z direction. The top layer is very high quality defect-free silicone nitrate, which is much higher quality than silicone dioxide. The lack of defects, as well as the gold or platinum metalization, allows the chemical electronics chip to function in harsh environments.

As features on chips have become smaller, the wafers from which the chips are made have become larger. Because of these trends, the silicon foundries for fabrication of chemical electronics are scarce. It is difficult to find a facility that has the capability to complete the entire device process. However, to continue with development of the integrated chemical sensing field, these facilities are a necessity.

The protection of closed spaces is another area where chemical sensors are of great importance. This concern is very closely related to transportation, since there is a high density of people in closed spaces on airplanes, trains, ships, subways, and other means of transit.

14

Future Challenges for the Chemical Sciences in Energy and Transportation

Safe, secure, clean, and affordable energy and transportation are essential to the economic vitality of the world. As we look to the future—the next 50 years and beyond—there will be many severe challenges to both energy and transportation created by population growth, economic growth, ever-tightening environmental constraints, increasing climate change issues and pressure for carbon dioxide emission limits, geopolitical impacts on energy availability and the energy marketplace, and a changing energy resource base. Science and technology—specifically the chemical sciences—will play a significant role enabling the world to meet these challenges. The opportunities are challenging and exciting. Advances in nanosciences, information sciences, biosciences, materials science, and chemical sciences will lead to solutions not contemplated today. The key will be fundamental research at the intersection of these sciences and developing new engineering to bring the new technologies to fruition.

To define the energy and transportation challenges and opportunities for the chemical sciences in the 21st century, future needs can be divided into two time frames—midterm (through 2025) and long term (2050 and beyond).[1] In the midterm:

- World energy demand will increase approximately 50 percent above 2002 levels. (Alexis Bell)
- Fossil fuels will remain abundant and available as well as continue to provide most of the world's energy. (Nathan Lewis)

---

[1] These future needs were identified by the committee based on the Workshop presentations. For each the presentation from which the need was identified is identified in parentheses.

- There will be a drive toward fuels with higher hydrogen-to-carbon ratio but balanced against the need to utilize the extensive low hydrogen-to-carbon coal resource base in the United States. (Venki Raman, Nathan Lewis)
- Tighter environmental constraints will be imposed. (Nathan Lewis)
- Government-mandated carbon dioxide limits will be initiated, leading to a need for carbon dioxide sequestration technology and introduction of large amounts of carbon-neutral energy. (Stephen Pacala)
- A real but acreage-limited role will be found for wind and hydro energy sources. (Nathan Lewis)
- Nuclear, solar, and biomass energy will play a growing role in the nation's energy mix. (Patricia Baisden, Jiri Janata)
- Cost-effective Hydrogen-2 fuel cell technology for transportation and power will be developed. (John Wallace, James Katzer)
- A significant penetration of vehicles with new high-efficient clear power sources will be seen in the transportation market. (John Wallace, James Katzer)
- Most Hydrogen-2 will be produced from fossil fuels.[2]

In the long term:

- World energy demand will rise to approximately two times the present energy usage. (Nathan Lewis)
- Fossil fuels will remain abundant and available, but limitations on their use will arise because of worldwide constraints on carbon dioxide emissions. (Nathan Lewis, Alexis Bell)
- There will be a need for significant carbon-neutral energy. (Most of the presenters)
- Fully developed carbon dioxide sequestration technology will be one of the important approaches to solving the energy problem. (Stephen Pacala)
- Coal and nuclear energy will continue to play a significant role in meeting world power demands. (Nathan Lewis, Alexis Bell)
- Renewable energy (wind, biomass, geothermal, photovoltaics, and direct photon conversion—e.g., solar photovoltaic water splitting) will play an increasingly important role. (Nathan Lewis, Ralph Overend)
- Most of world's vehicles will run on hydrogen from a carbon-free source or other fuels that are carbon-neutral. (John Wallace, James Katzer, Venki Raman)
- New cost-effective solar technology will be widely available. (Nathan Lewis, Ralph Overend)
- Hydrogen-2 and distributed electricity will be produced by solar energy, either through photovoltaic electrolysis or by direct solar photoelectrolysis. (Nathan Lewis, Ralph Overend)

---

[2]Venki Raman, in his presentation to the Energy & Transportation Workshop, noted that presently eighty percent of the hydrogen produced is made from natural gas steam methane generation.

While these scenarios can be debated, the drives they create in the chemical sciences are not greatly affected by the severity of the scenarios. They do point to a need to enhance the energy efficiency of fossil fuels in production and utilization, to develop a diverse set of new and carbon-neutral energy sources for the future, and to maintain a robust basic research program in the chemical sciences so that the technical breakthroughs will happen to enable this future.

The path will not be straightforward, however. While it is possible to predict research areas that most likely will have an impact on the development of more efficient energy and transportation systems, and direct resources to these research areas accordingly, looking back over the previous 50 years has shown that some of the most significant breakthroughs that have impacted energy and transportation were not foreseen. For example, advances such as the development of solid state physics and the broad applicability of lasers to many areas—not only in scientific research, but daily life as well—were not anticipated when these breakthroughs were first made. Advances such as these point to the continued importance of basic research. While future advances and challenges cannot always be predicted, robust long-term basic research can help to meet challenges, both anticipated and unexpected.

Particularly in the United States, interest and appreciation of the importance of science and technology is decreasing. Fewer U.S. students are entering technical careers. Energy research is decreasing significantly in both the private and public sectors. While this workshop and report do not address these issues, they must be resolved or the United States will be in jeopardy of not being able to meet its future energy and transportation requirements.

KEY CHALLENGES IDENTIFIED AT THE WORKSHOP ON ENERGY AND TRANSPORTATION

The needs of the energy and transportation sectors provide a number of challenges over the next century that the chemical sciences are uniquely suited to play a critical role. Many of the issues discussed in Workshop on Energy and Transportation, from increased energy efficiency from fossil fuels, to reduction of pollution, to sequestration of carbon dioxide, to development of new materials for vehicle fabrication, to new low cost renewable energy technologies, if not wholly chemical in nature, contain significant chemical science content. As chemical scientists seek to address these issues, the crosscutting nature of many of these challenges should be recognized at the outset. Many of the challenges in energy and transportation will be met with technologies that have broad applications in a number of different fields—new catalysts for increased reaction specificity and efficiency, new membranes for better separations, and new methods of fabrication to produce materials with predictable and very specific properties are just a few of many such examples. By working with scientists and engineers in other disciplines, such as materials scientists, bioscientists, geologists, electrical engineers, information scientists, mechanical engineers, and others, a multi-dimensional approach to these challenges will be accomplished—and the likelihood for comprehensive new solutions will increase significantly.

The following challenges were identified resulting from the presentations and discussions at the Workshop. Although these challenges were identified as a result of the Workshop, final responsibility for these statements rests with the organizing committee.

ENERGY

Fossil fuels will remain an abundant and affordable energy resource well into the 21st century. Since potential limitations on carbon dioxide emissions may restrict their utilization in the long term, it is imperative that chemical sciences research and engineering focus on making significant increases in the energy efficiency and chemical specificity of fossil fuel utilization.[3]

Professor Bell identified new multifunctional highly selective catalysts and membranes and corresponding process technologies as key research areas where opportunities will exist for major steps forward. These new catalysts and materials will allow much greater process efficiency (reduced carbon dioxide) through operations at lower temperatures and pressures and also by combining multiple process functions (i.e., shape selectivity and oxidation) in a single catalyst particle, thus reducing the number of process units in a plant.

These new materials and processes will increase the efficiency and environmental cleanliness of hydrocarbon production and refining and also enable refineries to produce chemically designed fuels for future vehicle power trains. These chemically designed fuels will play a key role in new power trains. These engines will require fuels that can optimize the efficiency of the entire power cycle while at the same time produce essentially no harmful exhaust. The best way to accomplish this is by designing the engine and fuel interactively, and this will lead to more chemical specificity requirements on the fuel.

Natural gas has tremendous potential for meeting the energy needs of the future because it has a high hydrogen-to-carbon ratio and can be converted to Hydrogen-2 and environmentally clean liquid fuels.[4]

Current technology for converting natural gas to liquid fuels is by Fischer-Tropsch technology, which converts methane to syngas (carbon monoxide and

---

[3]Alexis T. Bell, University of California, Berkeley, Presentation at the Workshop on Energy and Transportation.

[4]Alexis T. Bell, University of California, Berkeley, Nathan Lewis, California Institute of Technology, presentations at the Workshop on Energy and Transportation.

H_2) and the syngas to liquids (the Fischer-Tropsch step). While there have been major advances in the technology in the past decade, it is much less energy efficient than today's refining processes. New catalysts, membranes and processes are needed that will convert methane directly to H_2 and liquid fuels without going through syngas. This would tremendously increase the energy efficiency of methane conversion. Liquid products from these processes are chemically pure, containing no heteroatoms (i.e., sulphur, nitrogen, metals).

Management of atmospheric carbon dioxide levels will require sequestration of carbon dioxide. Research and development into methods to cost effectively capture and geologically sequester carbon dioxide is required in the next 10 to 20 years.[5]

As noted in Professor Pacala's presentation, effective management of the increasing anthropogenic output of carbon dioxide into the atmosphere will be a significant challenge for the chemical sciences and engineering over the 21st century. Development of sequestration technology to address this issue will require a thorough understanding of carbon dioxide chemistry and geochemistry along with an elaboration of the mechanisms involved in carbon dioxide absorption, adsorption, and gas separation. Also, effective sequestration will require new engineering knowledge to capture and transport the carbon dioxide to the sequestration site—most likely a geological reservoir. A more thorough understanding of the geochemical, geological, and geophysical nature of the sequestration site will be required to ensure that carbon dioxide does not escape over centuries of storage.

Biomass has the potential to provide appreciable levels of fuels and electric power, but an exceptionally large increase in field efficiency[6] is needed to realize the huge potential of energy from biomass.[7]

Biologically based strategies for providing renewable energy can be grouped into two major categories: (1) those that use features of biological systems to convert sunlight into useful forms (e.g., power, fuels) but do not involve whole living plants, and (2) those involving growth of plants and processing of plant components into fuels and/or power. Both are very important. Long-term improvements can be expected in the development of both biomass resources and the conversion technologies required to produce electric power, fuels, chemicals, materials, and other bio-based products. As molecular genetics matures over the

---

[5]Stephen W. Pacala, Princeton University, presentation at the Workshop on Energy and Transportation.

[6]In agriculture, field efficiency is the ratio of effective field capacity and theoretical field capacity.

[7]Alexis T. Bell, University of California, Berkeley, Nathan Lewis, California Institute of Technology, Ralph P. Overend, National Renewable Energy Laboratory, presentations at the Workshop on Energy and Transportation.

next several decades, for example, its application to biomass energy resources can be expected to significantly improve the economics of all forms of bio-energy. Improvements in economics, in turn, will likely lead to increased efforts to develop new technologies for the integrated production of ethanol, electricity, and chemical products from specialized biomass resources. Near-term markets exist for corn-ethanol and the co-firing of coal-fired power plants.

By the middle of the 21st century, global energy consumption will more than double from the present rate. To meet this demand under potential worldwide limits on carbon dioxide emissions, cost-effective solar energy must be developed.[8]

At present consumption levels, the supply of carbon-based fuels will be sufficient to meet our energy needs for well over a century. However, as noted in both Professor Bell's and Professor Lewis' presentations, the anticipated growth in energy demand over the next century, combined with climate change concerns, will drive the increased use of alternative sources of carbon-neutral energy. While a number of potential sources of renewable energy show promise for meeting part of this increased demand, including wind, biomass, geothermal, and expanded use of hydroelectric sources, solar power is most likely to meet the largest portion of this need. However, in order for use of solar power to increase substantially over the 21st century, new discoveries in photovoltaic and photochemical energy technologies must be made to reduce costs, increase conversion efficiency, and extend operating life. Advanced materials such as organic semiconductors and semiconducting polymers are needed to reduce energy costs from photovoltaics and make them competitive for electric power and H_2 generation. Current silicon-based photovoltaics are highly efficient but also very expensive. New technologies are needed to bring costs down. New photovoltaic materials and structures with very low cost-to-efficiency ratios—by lowering costs of fabrication, improving the efficiency, or both—will produce a step change in the use of photovoltaic technologies. For example, the use of grain boundary passivation with polycrystalline semiconductor materials might lead to the replacement of expensive single-crystal-based technology. The development of new, inexpensive, and durable materials for photoelectrochemical systems for direct production of hydrogen and electricity generation will be one of the main factors that will enable broad application of solar power to meet future energy needs.

Widespread use of new, renewable, carbon-neutral energy sources will require major breakthroughs in energy storage technologies.[9]

---

[8]Alexis T. Bell, University of California, Berkeley, Nathan Lewis, California Institute of Technology, Ralph P. Overend, National Renewable Energy Laboratory, presentations at the Workshop on Energy and Transportation.

[9]Nathan Lewis, California Institute of Technology, Henry S. White, University of Utah, presentations to the Workshop on Energy and Transportation.

Development of these technologies is dependent, in part, on breakthroughs in the design of energy storage systems due to the intermittent nature of many forms of renewables. Batteries, whose basic design has remained relatively unchanged for over a century, need to be fundamentally reexamined, as they will play an important role in meeting future energy needs. For example, Professor White highlighted advances in nanotechnology and its use in three-dimensional electrochemical cells as offering the possibility of increased energy density compared to conventional batteries, but these advances are still in the early stages of development. In addition, fundamental research breakthroughs are needed on thin-film electrolytes in order to develop high-power-density batteries and fuel cells.[10]

For full public acceptance of nuclear power, issues such as waste disposal, reactor safety, economics, and nonproliferation must be addressed.[11]

Future energy consumption trends indicate the need for additional sources of carbon-neutral energy. No one source of power will be sufficient to meet all of the projected increase in future power needs. Dr. Baisden in her presentation noted that nuclear power offers a plentiful supply of energy that is free from local emissions and produces no carbon-based greenhouse gases. However, nuclear power is unique in that political considerations are as important as technical challenges. One of the main technical challenges is waste management and disposal. Significant amounts of uranium can be reprocessed and reused in reactors, but this technology comes with significant concerns about nuclear proliferation and safety. Particularly in light of recent terrorist actions in the United States, the development of safe nuclear waste forms that not only will survive long-term repository storage but also allow secure transit to a repository remains an important priority.

Another significant issue facing the United States is the growing shortage of nuclear technical expertise. This threatens the management of the nation's currently installed nuclear capacity and certainly the development of the science and engineering needed to expand nuclear energy use in the future. The training situation is dire in nuclear chemistry, radiochemistry, and nuclear engineering. To address this shortage reinvestment in the education system will be required.

TRANSPORTATION

Vehicle mass reduction, changes in basic vehicle architecture, and improvements in power trains are key to improved vehicle efficiency. The

---

[10] At present fuel cell systems are being piloted for distributed generation backup power. This may provide another source of energy storage.

[11] Patricia A. Baisden, Lawrence Livermore National Laboratory, Jiri Janata, Georgia Institute of Technology, presentations to the Workshop on Energy and Transportation.

development and use of new materials are crucial to improved fuel efficiency.[12]

Dr. Sachdev noted in his presentation that reductions in the body mass of passenger vehicles will depend to a great extent on the successful integration of new light weight materials. The dual needs in these applications—for materials that are both lightweight and strong—continue to present challenges and opportunities in the chemical sciences.

The development of new polymers and nanocomposite materials will play an increasing role in vehicle mass reduction. The combination of high strength and light weight makes them ideal for many of these applications. Along with new materials, manufacturing and recycling processes will have to be developed that are both cost effective and environmentally effective.

As with the development of new catalysts, effective new materials benefit from a thorough understanding of structure/property relationships. This involves multiscale modeling and experimental efforts in surface science, including morphology. Enabling the use of new materials will also require extensive development of new nano- and microfabrication techniques, including biodirected or self-assembly syntheses.

Cost remains one of the main factors that determine both the need and the acceptance of new materials for applications in energy and transportation. In addition, passenger safety, which may be affected by the development of more lightweight vehicles, must also be taken into consideration. The imperative of low-cost, high-performance materials in the automotive industry will be driven by future environmental and CAFE regulations.

Reduced material cost is key to widespread use of the proton exchange membrane (PEM) fuel cell.[13]

As with other materials challenges, selective and energy-efficient separations are a highly desirable characteristic in many areas of energy and transportation research and engineering. Development of low-temperature, corrosion-resistant, thin membranes will further PEM development. However, development of new catalytic materials to replace the very expensive platinum in today's design is the most critical need.[14] Low-cost materials in fuel cells will be one of the key deciding factors in whether the United States readily transitions to a hydrogen economy.

---

[12]James R. Katzer, ExxonMobil, Kathleen C. Taylor and Anil Sachdev, General Motors Corporation, presentations to the Workshop on Energy and Transportation.

[13]John R. Wallace, Ford Motor Company, presentation to the Workshop on Energy and Transportation.

[14]A complementary goal to replacing expensive Pt in today's design is to develop catalysts with reduced Pt loading.

The lack of hydrogen generation, transportation, and storage infrastructure presents one of the main challenges to introducing hydrogen into the mass market as a transportation fuel and energy carrier.[15]

Effective hydrogen management and creation of the needed infrastructure will both be key to widespread adoption of hydrogen fuel cells to meet the country's energy needs for transportation and power. The challenges are great. New-generation technology is needed in the short- to midterm for hydrocarbon-based local refueling sites. In the long term, science and technology will have to be developed to generate hydrogen from carbon-free sources such as water, or at a minimum from carbon neutral sources. Whether this new energy source is based on nuclear, solar, or something that remains undiscovered, it will be one of the largest technical challenges the chemical sciences has ever undertaken.

Another significant challenge to effective hydrogen management is the development of efficient hydrogen storage, both onboard the vehicle and at a hydrogen generation facility. As with many other challenges, effective hydrogen storage is a crosscutting one that will require breakthroughs in a number of research areas. Progress is being made with metal hydrides and carbon nanotubes—but a commercial solution is a long way off. New materials will be key.

These technical challenges regarding hydrogen presently hinder widespread commercial use of hydrogen fuel cell technology for transportation and power. Until these challenges are met, it is unlikely that fuel-cell-powered vehicles will ever make up a significant portion of the passenger vehicle market.

CROSSCUTTING

Development of new, less expensive, more selective chemical catalysts is essential to achieving many challenges in both energy and transportation.

Catalysts are expected to play a role in virtually every challenge where chemical transformations are a key component. The development of new catalysts to solve challenges in energy and transportation will require the ability to design catalysts for specific needs. Utilization of new materials, nanotechnology, new analytical tools, and advanced understanding of structure/property relationships will create major catalytic advances. One of the major areas where these advances are needed is in controlling nitrogen oxide emissions from lean-burn engines and nitrogen oxide from coal power plants. Others are increased energy efficiency of fossil fuel processes, delivery of chemically designed fuels to new vehicle power systems, and direct conversion of natural gas into liquid fuels and Hydrogen-2. Another is the discovery of less expensive catalysts for the electroreduction of

---

[15]Venki Raman, Air Products and Chemicals, presentation to the Workshop on Energy and Transportation.

oxygen and the oxidation of fuels that can play an important role in fuel cells. As noted earlier, the global supply of Pt is insufficient to support a fuel cell transportation fleet using known electrode technology. Catalysts for promoting oxygen and hydrogen evolution from water, are also important in the design of photoelectrochemical systems.

CONCLUSION

Chemical research is required for substantial breakthroughs in the areas of energy and transportation. For example, the discovery of new catalysts, materials, and photoelectrochemical systems will require fundamental research in chemistry. Many of the challenges described above will only be met by effective interaction of the chemical sciences with other disciplines. In light of this, it is important to maintain a comprehensive and integrated approach to meeting these challenges. Also, chemical scientists should interact with researchers in other disciplines during the early stages of research planning in order to set and maintain this integrated approach. While it is not possible for chemical scientists to have a comprehensive knowledge of other disciplines, it is necessary for those in the chemical sciences to have a broad understanding of the nature of the interface in order for its impact to be fully appreciated.

When working to address these challenges, chemical scientists must always be watchful for unintended consequences. The energy and transportation sectors, being so closely tied to environmental impacts, must be particularly aware of solutions that may carry potentially negative impacts. Finally, in addition to scientific concerns, social, political, and economic impacts must be taken into account when addressing these challenges. Public perception and acceptance are key to many developments in energy and transportation and, as a result, should be considered when chemical scientists attempt to meet these challenges.

Because this report is based on only a 2-day workshop, details of chemical science research and engineering programs need to be further developed. The workshop's organizing committee suggests that the National Research Council pursue development of these detailed programs because of the importance of energy and transportation to our nation.

Appendixes

A

Statement of Task

The Workshop on Energy and Transportation was one of six workshops held as part of 'Challenges for the Chemical Sciences in the 21st Century.' The workshop topics reflect areas of societal need—materials science and technology, energy and transportation, national security and homeland defense, health and medicine, information and communications, and environment. The charge for each workshop was to address the four themes of discovery, interfaces, challenges, and infrastructure as they relate to the workshop topic:

- Discovery—major discoveries or advances in the chemical sciences during the past several decades.
- Interfaces—interfaces that exist between chemistry/chemical engineering and such areas as biology, environmental science, materials science, medicine, and physics.
- Challenges—the grand challenges that exist in the chemical sciences today.
- Infrastructure—infrastructure that will be required to allow the potential of future advances in the chemical sciences to be realized.

B

Biographies of the Organizing Committee Members

Allen J. Bard (NAS) (Co-chair) holds the Hackerman-Welch Regent Chair in Chemistry at the University of Texas, Austin. He received his B.S. from City College in 1955 and his M.A. and Ph.D. from Harvard University in 1958. His research focuses on electroanalytical and physical chemistry. Dr. Bard has been co-chair of the Board on Chemical Sciences and Technology and associate editor and editor-in-chief of the *Journal of the American Chemical Society*. He has received over 20 major awards and named lectureships and is a member of the National Academy of Sciences, American Academy of Arts and Sciences, American Chemical Society, and American Association for the Advancement of Science, and is a fellow of the Electrochemistry Society.

Michael P. Ramage (NAE) is retired Executive Vice President, ExxonMobil Research and Engineering Company. Previously he was Executive Vice President and Chief Technology Officer, Mobil Oil Corporation. Dr. Ramage held a number of positions at Mobil including Research Associate, Manager of Process Research and Development, General Manager of Exploration and Producing Research, Vice President of Engineering, and President of Mobil Technology Company. He has broad experience in many aspects of the petroleum and chemical industries. He serves on a number of university visiting committees and is a member of the Government University Industrial Research Roundtable. He is a Director of the American Institute of Chemical Engineers and a member of several other professional organizations. Dr. Ramage is a member of the National Academy of Engineering and serves on the NAE Council. He has a B.S., M.S., and Ph.D. in chemical engineering from Purdue University.

Joseph G. Gordon II is manager of Material Sciences and Analysis at IBM's Almaden Research Center. He received his A.B. from Harvard College (1966) and his Ph.D. in inorganic chemistry from the Massachusetts Institute of Technology (1970). His research involves interfacial electrochemistry, inorganic chemistry, and analytical chemistry. He has been a member of the Board of Chemical Sciences and Technology and is a member of the American Chemical Society, Royal Chemical Society, National Organization of Black Chemists and Chemical Engineers, American Association for the Advancement of Science, Electrochemical Society, and the American Physical Society.

Arthur J. Nozik is a senior research fellow with the Basic Science Division of the National Renewable Energy Laboratory (NREL). He received his B.S.Ch. from Cornell University in 1959 and his M.S. in 1962 and his Ph.D. in 1967 in physical chemistry from Yale University. Since receiving his Ph.D., Dr. Nozik has worked at NRL, where he has conducted research in nanoscience, photoelectrochemistry, photocatalysis, and hydrogen energy systems. He has served on numerous scientific review panels and received several awards in solar energy research. He is a senior editor of the *Journal of Physical Chemistry,* a fellow of the American Physical Society, and a member of the American Chemical Society, the American Association for the Advancement of Science, the Materials Research Society, the Society of Photo Optical Instrument Engineers, and the Electrochemical Society.

Richard R. Schrock (NAS) is Frederick G. Keyes Professor of Chemistry at the Massachusetts Institute of Technology. He obtained his B.A. in 1967 from the University of California at Riverside and his Ph.D. from Harvard University in 1971. In his research he first discovered and then focused on the preparation and properties of high-oxidation-state carbenes (alkylidine complexes). He also conducts research on the kinetics and mechanics of high-oxidation-state early-metal organometallic species and polymer synthesis and characterization as well as the synthesis of polymers containing organic or inorganic semiconductors or metal clusters. Dr. Schrock has been associate editor of *Organometallics,* has received several major awards, and is a member of the American Academy of Arts and Sciences and the National Academy of Sciences.

Ellen B. Stechel is on temporary assignment from her position as manager of the Chemistry and Environmental Sciences Department of the Ford Motor Company She received her A.B. in mathematics and chemistry from Oberlin College (1974) and her M.S. in physical chemistry (1976) and Ph.D. in chemical physics (1978) from the University of Chicago. She joined Sandia National Laboratories in 1991, where she became manager of the Advanced Materials and Device Sciences Department in 1994. Her scientific interests are in computational, surface, and materials sciences. Dr. Stechel serves as a senior editor for the *Journal of Physical*

Chemistry and on other editorial advisory boards. She has held numerous professional society positions with the American Vacuum Society, American Physical Society, American Chemical Society, and others. She also serves on U.S. Department of Energy committees.

C

Workshop Participants

**Challenges for the Chemical Sciences in the 21st Century:
Workshop on Energy and Transportation
January 7-9, 2002**

Alivisatos, Paul A., University of California, Berkeley
Alworth, William, Tulane University
Baisden, Patricia A., Lawrence Livermore National Laboratory
Baker, Thomas R., Los Alamos National Laboratory
Bard, Allen J., University of Texas, Austin
Barteau, Mark, University of Delaware
Bell, Alexis T., University of California, Berkeley
Bergman, Robert, University of California, Berkeley
Boron, David, U.S. Department of Energy
Bose, Rathindra, Kent State University
Bower, Stanley, National Renewable Energy Laboratory
Breslow, Ronald, Columbia University
Brinn, Ira, Georgetown University
Brown, Richard, University of Rhode Island
Carberry, John, DuPont Company
Card, Robert, U.S. Department of Energy
Carlin, Richard T., Office of Naval Research
Chuang, Steven, University of Akron
Clark, Sue, Washington State University
Cobb, James T., University of Pittsburgh
Cornell, Marty C., Dow Automotive

Delgass, Nicholas W., Purdue University
Devore, David, Ciba
Doctor, Richard, Argonne National Laboratory
Duraj, Stan, Cleveland State University
Erbach, Donald C., United States Department of Agriculture
Girolami, Gregory, University of Illinois
Gordon, Joseph, IBM
Gorte, Raymond J., University of Pennsylvania
Gottesfeld, Shimshon, MTI Microfuel Cell
Harrup, Mason, K., INEEL
Hrbek, Jan, Brookhaven National Laboratory
Janata, Jiri, Georgia Tech University
Johnson, Marvin, Phillips Petroleum Company
Kaldor, Andrew, ExxonMobil
Katzer, James R., ExxonMobil
Kirchoff, Bill, U.S. Department of Energy
Klein, Michael, State University of New Jersey
Knopf, Carl F., Louisiana State University
Kolb, Charles, Aerodyne Research
Kohl, Paul A., Georgia Institute of Technology
Kugler, Edwin, West Virginia University
Lamola, Angelo, Rohm & Haas
Laufer, Al, U.S. Department of Energy
Lewis, Nathan S., California Institute of Technology
Lo, Sunny, Dow Corning Corporation
Lochhead, Robert Y., The University of Southern Mississippi
Mallouk, Thomas E., Pennsylvania State University
Martin, Steve J., Dow Chemical Company
McIver, Robert T., IonSpec Corporation
Mullins, Michael E., Michigan Technological University
Murray, Royce, University of North Carolina Chapel Hill
Nguyen, Trung Van, University of Kansas
Nicholson, William J., Potlatch Corporation
Noble, Richard D., University of Colorado
Nocera, Daniel G., Massachusetts Institute of Technology
Nowak, Robert, Defense Advanced Research Project Agency
Nozik, Arthur, National Renewable Energy Laboratory
Oleson, John, OCI
Ou, John D., ExxonMobil Chemical
Overend, Ralph, National Renewable Energy Laboratory
Pacala, Steven W., Princeton University
Poeppelmeier, Kenneth, Northwestern University
Ponnampalam, Elankovan, MBI International

Powell, Joseph, Shell Chemicals, Limited
Rakestraw, Julie, DuPont Company
Ramage, Michael P., ExxonMobil (retired)
Raman, Venki, Air Products and Chemicals, Inc.
Rillema, Paul D., Wichita State University
Robinson, Sharon M., Oak Ridge National Laboratory
Rogers, Robin D., University of Alabama
Rogers, William, Pacific Northwest National Laboratory
Sachdev, Anil, General Motors
Saltsburg, Howard, Tufts University
Sattleberger, Alfred, Los Alamos National Laboratory
Schiehing, Paul, U.S. Department of Energy
Schobert, Harold, Pennsylvania State University
Schrock, Richard, Massachusetts Institute of Technology
Schuster, Darlene, American Institute of Chemical Engineers
Sciance, C. Thomas, Sciance Consulting Services, Inc.
Scouten, Charles G., Fusfield Group
Seltzer, Raymond, Ciba Specialty Chemicals
Semerjian, Hratch, National Institute of Standards and Technology
Sen, Ayusman, Pennsylvania State University
Shapiro, Pamela, University of Idaho
Siirola, Jeffrey, Eastman Chemical Company
Solomon, Jack, Praxair, Inc.
Stechel, Ellen, Ford Motor Company
Sutterfield, Dexter, U.S. Department of Energy
Tirrell, Matthew V., University of California, Santa Barbara
Tong, YuYe, Georgetown University
Turner, John, National Renewable Energy Laboratory
Utiedt, Roger, Minnesota Corn Processor
Valentine, Brian, U.S. Department of Energy
Van Zee, John W., University of South Carolina
Varjian, Richard, Dow Chemical Company
Venter, Jeremy, Rohm & Haas Company
Wallace, John R., Ford Motor Company
Warren, Timothy, Georgetown University
Wellek, Robert, National Science Foundation
White, S. Henry, University of Utah
Winkler, Philip W., Air Products and Chemicals
Wise, Jack, Consultant
Yow, Jesse, Lawrence Livermore National Laboratory
Zhao, Xinjin, W.R. Grace & Company
Zolandz, Raymond R., DuPont Company
Zoski, Cynthia, Georgia State University

D

Workshop Agenda

Challenges for the Chemical Sciences in the 21st Century
Workshop on Energy and Transportation
Lecture Room
National Academy of Sciences Building
2101 Constitution Avenue, N.W.
Washington, D.C.

Monday, January 7

    7:30 Breakfast
SESSION 1: CONTEXT AND OVERVIEW
    8:00 Introductory remarks by organizers. Background of project.
    8:00 **DOUGLAS J. RABER**, National Research Council
    8:05 **RONALD BRESLOW AND MATTHEW V. TIRRELL**, Co-Chairs, Steering Committee on Challenges for the Chemical Sciences in the 21st Century
    8:20 **ALLEN J. BARD**, Co-Chair, Energy and Transportation Workshop Committee
    8:30 **ALEXIS T. BELL**, *University of California, Berkeley*
           Research Opportunities and Challenges in the Energy Sector
    9:15 **KATHLEEN C. TAYLOR AND ANIL SACHDEV**, *General Motors*
           Materials Technologies for Future Vehicles
    9:45 Break
   10:15 **NATHAN S. LEWIS**, *California Institute of Technology*
           R&D Challenges in the Chemical Sciences to Enable Widespread Utilization of Renewable Energy

11:00 **STEPHEN W. PACALA,** *Princeton University*
Could Carbon Sequestration Solve the Problem of Global Warming?
11:45 General discussion
12:00 Lunch
SESSION 2: DISCOVERY
1:00 **R. THOMAS BAKER,** *Los Alamos National Laboratory*
Opportunities for Catalysis Research in Energy and Transportation
1:45 **HENRY S. WHITE,** *University of Utah*
Nano- and Micro-scale Electrochemical Approaches to Energy Storage and Corrosion
2:30 BREAKOUT SESSIONS
Breakout questions: What major discoveries or advances related to energy or transportation have been made in the chemical sciences during the past several decades? What is the length of time for them to show impact? What are the societal benefits of research in the chemical sciences? What are the intangible benefits, for example, in health and quality of life? What problems exist in the chemical sciences? Has there been a real or sustained decline in research investment in either the public or private sector? Has there been a shift in off-shore investment?
3:45 Break
4:00 Reports from breakout sessions (and discussion)
5:00 Reception
6:00 BANQUET –**DINNER SPEAKER: JACK SOLOMON,** *Praxair, Inc.*
The Chemical Enterprise and Vision 2020

Tuesday, January 8

7:30 Breakfast
SESSION 3: INTERFACES
8:00 **JAMES R. KATZER,** *ExxonMobil*
Interface Challenges and Opportunities in Energy and Transportation
8:45 **JOHN R. WALLACE,** *Ford Motor Company*
Fuel Cell Development—Managing the Interfaces
9:30 Breakout sessions
Breakout questions: What are the major discoveries and challenges related to energy and transportation at the interfaces between chemistry/chemical engineering and such areas as biology, environmental science, materials science, medicine, and physics? How broad is the scope of the chemical sciences in this area? How has research in the chemical sciences been influenced by advances in other areas, such as biology, materials, and physics?

10:45 BREAK
11:00 Reports from breakout sessions (and discussion)
12:00 LUNCH
SESSION 4: CHALLENGES
1:00 JIRI JANATA, *Georgia Institute of Technology*
Role of 21st Century Chemistry in Transportation and Energy
1:45 RALPH P. OVEREND, *National Renewable Energy Laboratory*
Challenges for the Chemical Sciences in the 21st Century
2:30 Breakout sessions
Breakout questions: What are the energy- or transportation-related grand challenges in the chemical sciences and engineering? How will advances at the interfaces create new challenges in the core sciences?
3:45 BREAK
4:00 Reports from breakout sessions and discussion
5:30 RECEPTION

Wednesday, January 9

7:30 Breakfast
SESSION 5: INFRASTRUCTURE
8:00 PATRICIA A. BAISDEN, *Lawrence Livermore National Laboratory*
A Renaissance for Nuclear Power?
8:45 VENKI RAMAN, *Air Products and Chemicals*
The Hydrogen Fuel Infrastructure for Fuel Cell Vehicles
9:30 Breakout sessions
Breakout questions: What are the energy or transportation-related issues in the chemical sciences, and what opportunities and needs exist for integrating research and teaching, broadening the participation of underrepresented groups, improving the infrastructure for research and education, and demonstrating the value of these activities to society? What returns can be expected on investment in the chemical sciences? How does the investment correlate with scientific and economic progress? What feedback exists between chemical industry and university research in the chemical sciences? What are the effects of university research on industrial competitiveness, maintaining a technical work force, and developing new industrial growth (e.g., in polymers, materials, or biotechnology)? Are there examples of lost opportunities in the chemical sciences that can be attributed to failure to invest in research?
10:45 BREAK
11:00 Reports from breakout sessions (and discussion)

APPENDIX D   103

   12:00   Wrap-up and closing remarks
            MICHAEL P. RAMAGE, Co-chair, Energy and Transportation Workshop Committee
   12:15   Adjourn

EXECUTIVE SESSION OF ORGANIZING COMMITTEE
   12:15   Working lunch: general discussion
    1:00   Develop consensus findings
    1:45   Develop consensus recommendations
    2:30   Develop action items, follow-up steps, and assignments
    3:30   Adjourn

E

Results from Breakout Sessions

A key component of the Workshop on Energy and Transportation was the breakout sessions that allowed for individual input by Workshop participants on questions and issues brought up during the presentations and discussions. Each color-coded breakout group (red, blue, green, and yellow) was assigned the same set of questions as the basis for discussions. The answers to these questions became the basis for the data generated in the breakout sessions. After generating a large amount of suggestions and comments, the breakout groups attempted to organize and consolidate this information, sometimes voting to determine which topics the group decided were most important. After each breakout session, each group reported the results of its discussion to the entire workshop.

The workshop committee has attempted in this report to integrate the information gathered in the breakout sessions and to use it as the basis for the findings contained herein.

Red Team Challenges

Challenges

- Hydrogen storage
- Direct conversion of methane
- Hydrogen from thermochemical sources (hydrogen without carbon emission)
- Low-energy selective separation
- Carbon dioxide management
- High-power-density energy conversion devices

Enabling Techniques

- Low-cost catalysts
- Interface thermodynamics and kinetics
- Advanced membranes for gas separation
- Artificial photosynthesis
- Designer fuels
- Thin-film electrolytes
- Computation
- Detection
- Sensors

Blue Team Challenges

Catalysis (enabling science)

- Catalysis by design
- Catalysis for fuel cells
- Efficient carbon dioxide reduction to fuels
- Direct conversion of Methane

Photovoltaic and Photoelectrochemical Cells and Energy Storage

- High efficiency
- Low cost
- Storage

Separation Technology—the science behind

- High temperature polymeric membranes for fuel cells
- Selective separation propane/propylene

Conversion of Hydrocarbons to Oxygenated Fuels

- The fundamentals

Tie between:

> **Hydrogen Storage at Ambient Conditions**
> **Fundamentals-Based Computer Modeling of Reactions and Processing**

Better Processes

Define Objective in Energy and Transportation

- Quantities of energy saved
- Renewables introduced
- Carbon dioxide & environmental effects
- Future needs of power

Chemical Sciences Contributions to Above

Green Team Challenges

Predictable Chemical Catalysis (e.g., nonprecious metals)

Cheap Renewable Energy

- Solar electricity
- Hydrogen-2
- "Artificial photosynthesis"

In situ diagnostics for sensing and control on all timescales

Predictable Materials Design

- Low temperature ion conductors
- Mesoscale
- Interfaces and surfaces
- Noncorroding materials

Complete Elaboration of Carbon dioxide Chemistry

- Low-cost carbon sequestration
- Carbon dioxide activation

Yellow Team Challenges

Energy and Transportation that are Safe, Affordable, and Desirable (lean, green, and keen)

- More efficient and selective chemical conversion
- Transportation without harmful emissions
- Gas-to-liquid technology
- Cheap H_2
- "Smart" highways supplying energy
- Energy management in vehicles
- Energy storage

Development of Energy Systems that Are Secure and Sustainable

- Emissions management—eliminate nitrogen oxide and carbon dioxide
- Solving the greenhouse problem
- Managing nuclear waste
- Low-energy water purification
- Fixing atmospheric carbon dioxide: artificial photosynthesis
- Selective and energy-efficient chemical separations

Predictive Synthesis of Materials with Desired Properties and Functionality

- Cost-competitive materials for solar power
- Cost-competitive materials for fuel cells
- Cheap, durable, room-temperature superconductors
- Cost-effective high-performance materials—composites, light alloys, etc.
- Thermoelectrics with high ZT
- Molecular understanding of energy and molecular conversion

Yellow Team Interfaces

Fuel Cells (P.E.M.)

- Electrocatalysis (O_2 reduction)
- Bipolar plate design
- Enhanced electrode interface
- Fuel cell membranes

H_2 Storage

Carbon Dioxide Management

Water Purification/Selective Separations

Lean-Burn Emissions

Lightweight Materials

Sensor

Biocatalysis/Genetic Engineering (Hydrogen-2 production, ethanol, Sulphur-removal)

Ceramic Membranes/Solid Oxide Fuel Cells

Photovoltaics

Nuclear Life Cycle/Reactor Engineering

Green Team Interfaces

Divided by Grand Challenges (bullets in order of votes)

Predictable Catalysis

- Nano-microfabrication (physics, materials science interface with the chemical sciences)
- Structure/property relationships (physics, materials science)
- Biodirected/biocatalyzed synthesis (biology)

Cheap Renewable Energy: Photovoltaics, H_2, Artificial Photosynthesis

- Hydrogen-2 storage (mechanical engineering, materials science)
- Charge transport/optical properties, photophysics (electrical engineering, physics, materials science)
- Design of biomaterials/biomass (biology)
- Understanding bio-energy transduction (biology)
- Bacterial production of Hydrogen-2 (biology)
- Heavy metals management (ecology)

In situ Sensing and Control

- Analytical, optical, spectroscopic, solid-state sensors (electrical engineering, physics)
- Biosensors (biology, electrical engineering)

Predictable Materials Design

- Life-cycle assessment (ecology, material science)
- Multiscale modeling: molecular, nanoscale, mesoscale (physics, materials science, computer science, mathematics)
- Biodirected synthesis/self-assembly (biology, physics, materials science)
- Surface science: morphology, beam technology (physics, materials science)

Complete Elaboration of Carbon Dioxide Chemistry

- Carbon-cycle analysis (climatology, oceanography, geology)
- Sequestration (geology, climatology)
- Life-cycle analysis (ecology)
- Design of biomass (biology)

Red Team Interfaces

Comprehensive and Integrated Approach

Interact with all Interfaces in Early Stages of Research Planning

Understand Entire Interface for Full Impact

Look Out for Unintended Consequences

Materials Science (Really Is Chemistry)

- Thermochemical production requires new materials
- H_2 storage requires new materials (ambient/solid)
- Mesoscale structure and behavior
- Predictable performance
- New electrolytes

Information Technologies

- Modeling, analysis, complexity
- Data: storage, visualization, mining, fusion
- Collaborative tools/capabilities
- Mathematics of sensor arrays
- Signals and signal processing

Biomimetic Processes and Synthesis

- Direct conversion of methane
- Low temperature catalysis
- Efficiency and selectivity

Physics

- Solid-state devices
- Electrooptical
- Sensors

Mechanical Engineering

- Thermal and mechanical packaging
- Efficient energy-to-work conversion
- Fuel cell and battery packaging

External

- Social, political, and economics sciences
- Model science and social and economic factors
- Public perception and acceptance
- Consider multidimensional impacts
- Environment and health
- Predictive toxicology

Team Blue Interfaces

Solid-State Physics

- Semiconductors
- Sensors
- Transport and storage of H_2
- Catalysis

Computational Sciences

- Catalysis by design (e.g., single-site catalysis for polyolefin polymerization)

Materials

- Nanostructures
- Separations
- Hydrogen storage
- Photovoltaics

Biomimetics

- Catalysis
- Energy conversion
- Biosensors